METAL-CERAMIC JOINING

METAL-CERAMIC JOINING

Proceedings of a symposium held on Metal-Ceramic Joining, sponsored by the TMS Refractory Metals Committee, held during the TMS Fall Meeting, Detroit, Michigan, October 8-9, 1990.

Edited by

P. Kumar
Cabot Corporation
Boyertown, PA 19512

and

V. A. Greenhut
Center for Ceramic Research
Rutgers, The State University of New Jersey
Piscataway, NJ 08855-0909

A Publication of

TMS
Minerals • Metals • Materials

A Publication of The Minerals, Metals & Materials Society
420 Commonwealth Drive
Warrendale, Pennsylvania 15086
(412) 776-9024

Printed in the United States of America
Library of Congress Catalog Number 91-61798
ISBN Number 0-87339-128-4

Preface

A Symposium on "Metal-Ceramic Joining" was held during the Metals Congress in October 1990. It was sponsored by the Refractory Metals Committee of TMS. Most of the papers which were presented in this symposium are complied in this book. About two-third of the papers are related to the high-temperature applications; the remaining papers pertain to the electronic applications.

Cawley's paper is designed to serve as an introduction to some of the central issues associated with metal-to-ceramic joining. As such it is complementary to the overview of joining techniques, and is intended to serve as an introduction to the broad range of focussed work, reported in this volume.

Kaysser et al. calculated the high, failure-inducing residual stresses between a iron-nickel superalloy and silicon nitride, and compliant interlayers introduced to lower the thermoelastic residual stress. Hot isostatic pressing (HIP) was employed to diffusion bond various materials. Iron alloys with good formability and low thermal expansion were used as intermediate layers. These were combined with vanadium, hafnium and niobium-based layers which had the purpose of promoting bonding. The reactions at layer interfaces were examined by SEM and X-ray fluorescent microanalysis.

Although it has been recognized for a long time that the cool down cycle from the brazing temperature can be very important for the achievement of low residual stresses in metal/ceramic brazes, no methodology until now has permitted simulation of those stresses. The paper by Stephens, et al. examined the incorporation of minimum creep rate correlations for Cu into FEA (Finite Element Analysis) codes, and indicated that quantitative computational treatment of this problem, although in its infancy, shows promise for use as a predictive design tool.

Peteves and Nicholas joined the silicon nitride to sialon ceramics using active metal brazing and active metal diffusion bonding. A variety of materials sources were examined. The active metals, titanium and chromium were included in the alloy and/ or vapor deposited upon the ceramic. The wetting response, bond strength and Weibull modulus were determined. The effect of various gas environments during bonding was examined. The structure and composition of the interfaces was examined by optical microscopy, SEM, and X-ray fluorescent microanalysis.

Majumdar and Ahmad also used active-filler metal approach to join cast iron to zirconia. The fracture toughness of the resulting bonds were used to confirm the theoretical model for a mixed-mode fracture in a ceramic-metal joint.

Greenhut gives an overview of the various techniques for joining and metallization of ceramics. He also discusses recent developments in these techniques.

Haber and Greenhut examined the bonding of aluminum to alumina using borosilicate glass in the ceramic composition. They found an optimum level of borosilicate glass addition for maximum strength.

In Levy's paper, the residual stress in brazed ceramic-metal joints was calculated and compared for a closed form model and iterative finite element analysis. The calculated stresses showed agreement with observed failure modes in active metal braze joints between alumina and silicon carbide ceramics with titanium, tungsten, tantalum and titanium aluminide. The microstructure and chemistry of joined interfaces was examined by optical microscopy and SEM/EDS analysis.

Holowczak et al. investigated the effect of sintering aids in alumina on the peel strength of copper-alumina bonds. They conclude that sintering aids significantly affect the peel strength; simultaneous additions of calcia and magnesia improve the bond strength.

Park et al. conducted the active metal brazing of nodular iron and magnesia-stabilized zirconia by predeposition of titanium on the ceramic substrate surface and use of a braze containing titanium active metal. Ultrasonic C-scan tests and direct examination of large scale joints showed voids and poor contact in the case of predeposition. Active filler metal showed superior contact and bonding. Sources for possible titanium oxidation were identified and examined using equilibrium thermodynamic calculations.

In Knapp's paper, alumina ceramics were glass bonded to each other employing a glass with the composition of anorthite. The authors term this technique "silicate brazing" in analogy to metallurgical methods. Bond strength was quite variable as determined by four point bending with an average strength of about 145 MPa. Partial crystallization of the glass bond layer deteriorated strength. Interfaces were characterized for microstructural and elemental distribution by optical microscopy and electron microprobe analysis.

Carpenter of NIST discusses the measurements of a variety of properties of the ceramic-metal bonds in microelectronics and multilayer capacitors. These measurements are required in order to understand local and interfacial properties in a device. In-service failures, design and new applications require such measurements. The increased spatial density of these features requires further development of property measurement techniques. The present state of such techniques as well as possible future developments are discussed as they pertain to: dielectric properties, thermal diffusivity, thermal expansion, stress-strain behavior, adhesion, fracture toughness, residual stress and curing of polymeric materials.

C. H. Li discussed the graded seals for metal-ceramics joints. The thermal coefficient of expansion of these seals varies axially, which minimizes the dynamic mismatch stresses of the joint. These seals are particularly useful for mounting fragile semiconductors and superconductors to metal substrate for improved bond strength, heat dissipation and device performance.

Brow and Loehman investigated the effect of titanium additions to the eutectic braze compositions via the electron microscopy of the interface in metal - AlN system. Titanium reduces the contact angle and improves the wetting. In addition, the formation of TiN at the interface improves the bonding.

Typically, glass technologists and ceramists consider metals to be relatively inert; while metallurgists tend to have the same viewpoint with respect to glasses and ceramics. The fallacy of these viewpoints was exposed in the paper by Knorovsky and Brow looking at a glass ceramic to metal debonding problem between matched coefficient of expansion materials (304L SS and a lithium silicate-based glass ceramic). In this case, an integrated materials system approach was necessary to understand the problem (a phase transformation on the metal side of the interface driven by Cr depletion into the glass-ceramic) and look for solutions.

Chiang et al. investigated bonding of AlN-Cu system using gas-metal eutectic and solder methods. Several important factors were identified to play a major role on the bond-strength. High thermal conductivity of AlN makes it an attractive alternate to alumina in electronic packages.

We would like to thank the authors for sending manuscripts in on time. Catherine Yoder and Olympia McNicols have provided invaluable secretarial assistance.

P. Kumar
V. A. Greenhut

Table of Contents

I. High Temperature Applications

II. Joining Processes

III. Electronic Applications

High Temperature Applications

INTRODUCTION TO CERAMIC-METAL JOINING

James D. Cawley

Department of Materials Science and Engineering
The Ohio State University
2041 College Road
Columbus OH 43210-1178

Abstract

This paper is designed to serve as an introduction to some of the central issues associated with ceramic-to-metal joining. As such it is complementary to the overview of joining techniques, and is intended to serve as an introduction to the broad range of focussed work, reported in this volume.

The author gratefully acknowledges the support of the Edison Welding Institute Core Research Program for research work related to metal-ceramic joining.

Metal - Ceramic Joining
Edited by P. Kumar and V. A. Greenhut
The Minerals, Metals & Materials Society, 1991

Joining is a fundamental processing step and consequently has a long history as well as a continually expanding scientific base. Several comprehensive review articles [1-5], a conference proceedings [6] and a text [7] on ceramic joining have all been published in recent years. Rather than attempt a comprehensive overview in this manuscript, a selected set of illustrative examples is employed.

Definitions

Joining may be defined as the controlled formation of a bond between two constituents of a component after a desired geometrical geometrical arrangement has been established. A bond may, in turn, be defined as a low energy state, i.e. an energy penalty is associated with separating two bonded interfaces. It is traditional to discuss bonds of two types: primary (or chemical) and secondary (or physical). Both are usefully exploited in metal-ceramic joining.

The process of joining is simply to bring the adherends (or materials to be joined) into approximate geometry while in a transient high energy state and to allow the system to relax, or dissipate this energy. The driving force for bond formation may include several terms; for example, the reduction of surface energy and the relief of chemical potential gradients through diffusion and/or chemical reaction. The controlled application of heat (either local melting or isothermal annealing) may be used to determine the kinetics of bond formation. In some cases, chemical reactants are brought into contact just prior to bonding and the kinetics of the resultant reaction(s) control bond formation.

Joining may also be defined as a secondary processing step and as such the consequence of the the joining procedure on the microstructure, and any concomitant change in properties, must be considered. In general, joining processes and materials are selected which minimize microstructural changes. This issue is particularly acute when attempting to join materials which are characterized by different classes of interatomic bonding.

As a general rule, interatomic bonding in metals is primarily metallic or mixed metallic/covalent, in ceramics interatomic bonding may be considered either ionic, covalent or mixed ionic/covalent, and organic polymers are characterized by intramolecular covalent bonding and intermolecular hydrogen or van der Waals bonding. One advantage of classifying materials based on interatomic bonding (as opposed to, for example, the end use of the material such as structural materials, electrical materials, optical materials...) is that many of the properties relevant to the design of a joining process are strongly correlated to interatomic bonding, see Table I.

Joining Processes

A wide variety of techniques have been employed in ceramic-to-metal joining. These include: adhesive bonding; metal brazing; silicate brazing; and diffusion bonding. It is possible to identify similarities as well as differences between these processes. The first three all involve the use of a foreign interlayer material between the contacting surfaces of the adherends and therefore (at least) two qualitatively different interfaces and produced during the joining operation. The properties of the interlayers, however, are determined by secondary bonding in the case of adhesive bonds in contrast to chemical bonding in the case of brazing and diffusion bonding.

Table I. Material Properties Associated with Interatomic Bonding

Metallic Bonding
- Close packed crystal structures
- Small volume change upon solidification accommodated by dislocation motion
- High relative thermal expansion
- May have modest melting temperatures

Ionic Bonding
- Close packed crystal structures
- Limited dislocation mobility (localized charges)
- May have large solidification volume change
- Intermediate relative thermal expansion
- High melting temperatures (with multivalent ions)

Covalent Bonding
- Open crystal structures
- Limited dislocation mobility (directional bonds)
- May expand upon solidification
- Low relative thermal expansion
- May decompose or sublime

Secondary Bonding
- Amorphous or partial crystallization
- Low (several hundred degrees Celsius) melting points

Adhesive joints offer a great deal of flexibility. They may be formed at, or near, room temperature (which presumably will be the use temperature), have near zero residual stress, and distribute stresses during service. However, adhesive joints may be limited in terms of hermeticity, electrical properties, and high temperature service. Metallic brazing, silicate brazing, and diffusion bonding are all exploited in structural and electronic applications; e.g., active metal brazing of silicon nitride turborotors to steel shafts, solder glasses for electronic packaging, and diffusion bonding of pins to ceramic heat sinks for electronic applications.

These three procedures al involve heating dissimilar materials to high temperatures. As a result, the issues of interfacial chemistry and thermal stresses are central to the success of each type of joint.

Interfacial Chemistry

Wetting and spreading are important when a liquid phase is present at the contacting surfaces (i.e. brazing). For nonreactive systems this is typically analyzed using the Young-Dupré equation which relates the equilibrium angle of a sessile drop on a planar surface measured through the liquid, θ, to the surface tensions associated with the solid/vapor, solid/liquid and liquid/vapor interfaces (γ_{sv}, γ_{sl} and γ_{lv} respectively) and may be written [e.g. 7]

$$\cos\theta = (\gamma_{sv} - \gamma_{sl}) / \gamma_{lv}. \qquad \text{Eqn-1}$$

It is conventional to distinguish three regimes of behavior in terms of θ or $\cos\theta$:

Nonwetting	$\theta > 90°$	$\cos\theta < 0$
Wetting	$\theta < 90°$	$\cos\theta > 0$
Spreading	$\theta = 0°$	$\cos\theta = 1$.

It is evident that spreading is simply a special case of wetting. Since interfacial energy is always positive the sign of the right hand side of Eqn-1 is uniquely determined by the difference $\gamma_{sv} - \gamma_{sl}$. As long as the wetted surface is of lower energy than the exposed surface, wetting will result. The magnitude of the effect is controlled by the value of γ_{lv}. When $\gamma_{lv} >> (\gamma_{sv} - \gamma_{sl})$ then $\cos\theta \to 0$ (or $\theta \to 90°$) regardless of the sign of the numerator. On the other hand, when $\gamma_{lv} << (\gamma_{sv} - \gamma_{sl})$, $\cos\theta$ -> becomes a large positive number and spreading results ($\theta \to 0°$).

In most brazing operations wetting is required, but spreading is undesirable. Spreading may lead to excessive surface flow, deplete the joint region of the filler metal or cause sticking to the support structure or clamps, etc. This problem is schematically illustrated in Fig. 1. A neutral condition, i.e. no spreading or dewetting, is preferred when a foil of the braze metal is preplaced in between the contacting surfaces. Filling a gap through a capillary action does not require spreading either. The expression [8] for the velocity of the fluid front during capillary rise in between two plates, v, is

$$v = d\, \gamma_{lv} \cos\theta / (4\, \mu\, h)$$

where d is the separation between the plates, h is the rise after a given time and μ is the viscosity of the liquid. The equilibrium rise against gravity, H, (the maximum extent of penetration by the liquid) may be expressed as

$$H = 2\, \gamma_{lv} \cos\theta / (\rho\, d)$$

where ρ is the density of the liquid. Both quantities are dependent on the product $\gamma_{lv} \cos\theta$ which is equivalent to the difference $\gamma_{sv} - \gamma_{sl}$ and therefore independent of γ_{lv}. The physical significance of this is that capillary penetration does not create liquid surface exposed to the atmosphere. A high value for γ_{lv} may be beneficial, as it results in a large energy penalty for fluid flow out of the joint.

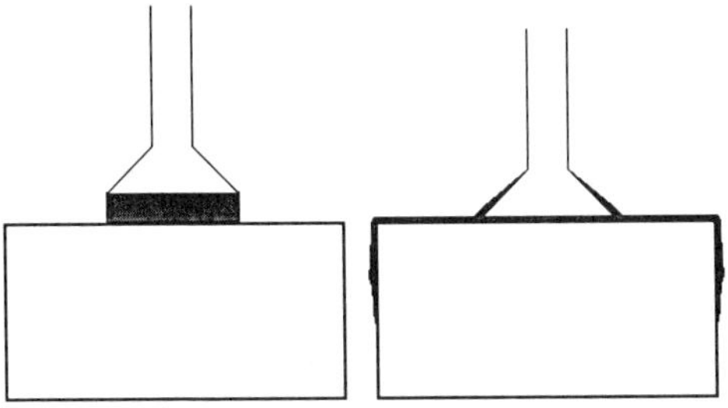

Fig. 1.The schematic on the left represents the behavior expected when the molten metal wets the contacting surface, but the interfacial tension associated with the liquid vapor interface is high. The figure on the right depicts the deleterious nature of excessive surface flow as a result of spreading, i.e. low liquid surface tension.

The low surface tension of most ceramics has the consequence of making it unlikely for many molten metals to satisfy the wetting condition. As an example, the interfacial energies for aluminum oxide and various nonreactive metals is shown in Table II. In addition, wetting alone is insufficient to assure the formation of a strong bond. Molten silicates may wet a metal surface, but not form chemical bonds at the interface due to the abrupt discontinuity in the type of interatomic bonding.

	Interfacial Energy (mJ/m^2)				
Material	γ_{sv}	γ_{sl}	γ_{lv}	θ (deg)	T(°C)
Al_2O_3	905	-	-	-	-
Cu	-	2245	1220	170	1100
Fe	-	2270	1720	128	1550
Fe-3.9C	-	1365	1720	105	1570
Ni	-	2205	1725	138	1475

B. C. Allen and W. D. Kingery, Trans. AIME, 215, p. 30-7 (1959).

All of the above discussion relates to material properties at or very near to the interface and therefore may be easily altered through the use of coatings produced either in a discrete processing step or in situ as the result of a chemical reaction. The goal is often expressed as grading the interatomic bonding near the interface to allow brazing between like-to-like surfaces. Either the surface of the ceramic or the metal may be altered. Metallization of a ceramic (e.g. the moly-mangenese process on debased alumina [8,9] or Cr-metallization of silicon carbide [10]) prior to metal brazing and oxidation of a metal prior to glass-to-metal sealing [8] are examples of the inclusion of a discrete processing step and active metal brazing [4,11] (which relies on segregation and reaction to form a thin coating on the ceramic) and the use of adherence oxides [3,12] in porcelain enameling are examples of techniques which exploit in situ reactions.

The exploitation of in situ reactions is desirable in that it simplifies the processing sequence, but the often the desired reaction products are metastable and aging of joint at elevated temperatures may lead to degradation of the joint.

The development of a strong bond at elevated temperature, either by brazing or diffusion bonding, does not assure good system performance. Reaction products at the interface may lead to embrittlement and thermal stresses which develop during cooling may be intolerable.

Thermal Stresses

Differential thermal strains during cooling are to be expected in general. To a first approximation, differential thermal strain, $\Delta\varepsilon$, may be evaluated by the mismatch in coefficients of thermal expansion (CTE), $\Delta\alpha$, by

$$\Delta\varepsilon = \Delta\alpha \, \Delta T \qquad \text{Eqn-2}$$

and, depending on the geometry of the specimens as well as the values of the relevant elastic constants and ductilities, the resultant state of stress may be analyzed. As illustrated in Fig. 2, the overlap between metallic systems and ionic ceramics is limited to refractory metals and there is almost no overlap with between metals and covalently bonded ceramics.

Although ceramics are relatively strongly when loaded in compression, it is not sufficient to simply choose a set of materials which should result in compressive stress in the ceramic since, depending on joint geometry, there may be localized regions of high tensile stress, as illustrated in Fig. 3. The formation of direct bonds between a metal and a ceramic therefore requires either an exceptionally good match in CTE or carefully designed joint geometry. The niobium-alumina system is an example of the former and Houskeeper seals [8] (in which feather edged copper is joined to glass of low relative CTE) is an example of the latter.

Even when two materials have similar net thermal strains when cooled from the bonding temperature to room temperature, nonlinear thermal expansion (e.g. glasses heated through the glass transition or as a result of displacive phase transformations) may result in significant transient stresses. This is schematically illustrated for the case of a metal and a silicate glass in Fig. 4. The local stresses may be reduced either through stress relaxation (e.g. viscous flow of a silicate glass or plastic flow of a ductile metal).

Grading of the CTE difference through the use of interlayers is a one standard approach to minimizing the development of residual stress. In one approach a thin slab(s) of a material(s), which remains solid throughout the joining process, having an intermediate CTE is placed between the adherends. The process, therefore, requires the production of two physically distinct joints. In some instances different processes (or braze metals) may be used at the two interfaces. An alternate procedure is to produce a composite interlayer which has an intermediate CTE. Usually this consists of a low expansion phase (e.g. carbon fibers) to be distributed within a high expansion matrix of a brazing alloy (e.g. Ag-Cu alloys). A possible complication associated with such a composite is that the inclusion of a high specific surface area nonoxide phase in a reactive brazing alloy may consume a large fraction of the active element and significantly alter the composition of the alloy.

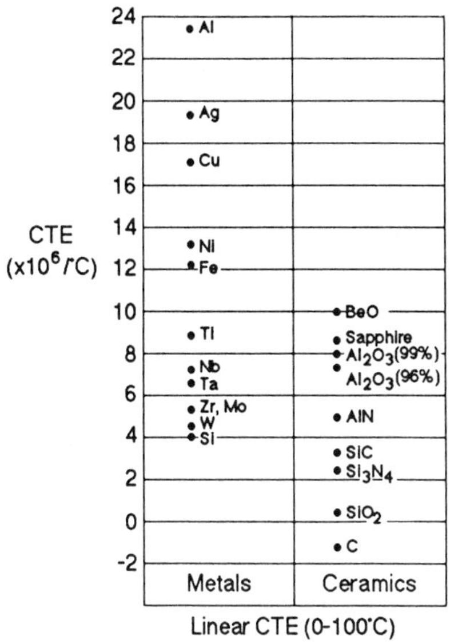

Fig. 2. Graphical depiction of the thermal expansion coefficients of selected ceramics and metals at low temperature. Reference to this figures indicates large thermal strains will be associated with brazing alloys based on the silver-copper eutectic and that covalently bonded ceramics are particularly difficult to match. (After T. Tanaka, H. Morimoto and H. Homma, Nippon Steel Tech. Rep. No 37, 1988, p. 35)

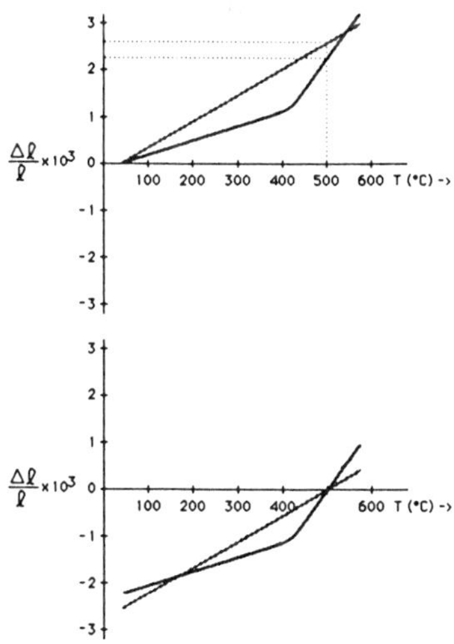

Fig. 3. Schematic illustration of differential thermal strain versus temperature associated with a glass-to-metal seal. The upper graph depicts thermal strains experienced upon heating (with the specimens unbonded and l_o is defined as l at room temperature) and the lower curves refer to shrinkage during cooling from an arbitrary bonding temperature of 500°C (in this case, l_o is defined as l at 500°C). Although the overall thermal strain from 500°C - RT is small and favors compression in the glass differential strain, a large transient tensile stress would be produced during the initial stages of cooling when the contraction rate of the glass is high.

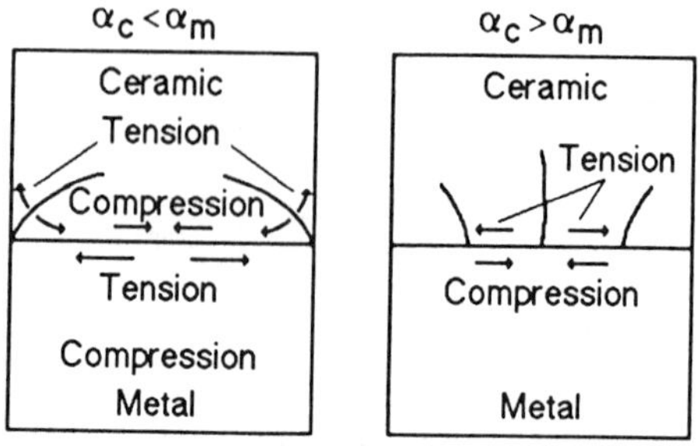

Fig. 4. Schematic illustration of the failure modes observed for systems with large CTE mismatch. (After T. Tanaka, H. Morimoto and H. Homma, Nippon Steel Tech. Rep. No 37, 1988, p. 36)

Summary

Joining processes of particular relevance to electronic and structure applications of ceramic-to-metal joints include diffusion bonding, metal brazing, and silicate brazing. Two generic issues associated with these processes are interfacial chemistry and thermal stress development. An introduction to these subjects has been provided.

References

1. R. Rice, "Joining of Ceramics," in Advances in Joining Technology, Edited by J. J. Burke, A. E. Gorum, and A. Tarpinian (Boston, MA: Brooke Hill, 1975), 69-111.

2. M. G. Nicholas and D. A. Mortimer, "Ceramic/Metal Joining for Structural Applications," Mater. Sci. Technol., 1 [9] (1985), 657-665.

3. J. A. Pask, "From Technology to the Science of Glass/Metal and Ceramic/Metal Sealing," Am. Ceram. Soc. Bull., 66 [11] (1987), 1587-1592.

4. R. E. Loehman and A. P. Tomsia, "Joining of Ceramics," Am. Ceram. Soc. Bull., 67 [2] (1988), 375-80.

5. J. D. Cawley, "Joining of Ceramic Matrix Composites," Am. Ceram. Soc. Bull., 68 [9] (1989), 1619-1623.

6. "Proceedings of the International Forum on Structural Ceramics Joining," Edited by R. E. Loehman, S. M. Johnson, and A. J. Moorhead, Ceramic Engineering and Science Proceedings 10 [11-12] (1989).

7. M. M. Schwartz, Ceramic Joining, (Materials Park, OH: ASM International, 1990).

8. W. H. Kohl, Handbook of Materials and Techniques for Vacuum Devices, (New York, NY: Reinhold Publishing Corp., 1967)

9. D. M. Mattox and H. D. Smith, "Role of Mangenese in the Metallization of High Alumina Ceramics," Am. Ceram. Soc. Bull., 64 [10] (1985), 1363-67.

10. H. Okamura, T. Shida, T. Funamoto, "Brazing of High Conductivity Ceramics," presented at the 5th International Symposium of the Japan Welding Society on on "Advanced Technology in Welding, Materials Processing and Evaluation," Makuhari, Tokyo, April 17-19, 1990.

11. H. Mizuhara, E. Huebel, and T. Oyama, "High Reliability Joining of Ceramic to Metal," Am. Ceram. Soc. Bull., 68 [9] (1989), 1591-1599.

12. K. A. Maskall and D. White, Vitreous Enameling: A Guide to Modern Enameling Practice, (New York, NY: Pergamon Press, 1986.

REACTIVE DIFFUSION BONDING

OF Si_3N_4 TO MA6000

W.A.Kaysser, A.Frisch, W.Zhang, and G.Petzow

Max-Planck-Institut für Metallforschung
Pulvermetallurgisches Laboratorium
D-7000 Stuttgart 80, Germany

Abstract

Diffusion bonding of Si_3N_4 to MA-6000 is a method considered to yield reliable joined metal/ceramic systems for high temperature applications. Joining of Si_3N_4 to MA-6000 bars was achieved by diffusion bonding during Hot Isostatic Pressing (HIP) at 100 to 200 MPa at 1100 to 1400°C. Stresses caused by the large difference in thermal expansion coefficients were reduced by multiphase interlayers. Iron-based alloys with high formability at elevated temperatures and small thermal expansion coefficients at low temperatures were combined with V, Nb and Hf-based layers. The reactions at the layer interfaces and the fracture surfaces were investigated by SEM, EDX and WDX analysis.

Metal - Ceramic Joining
Edited by P. Kumar and V. A. Greenhut
The Minerals, Metals & Materials Society, 1991

Introduction

Metal/ceramic bonds are required in the electronics field for insulators, capacitors or microchip substrates /1/ and as structural materials where high toughness and good resistance to wear, corrosion and erosion at high temperature operations /2/ are required. Si_3N_4 and Ni-based superalloys, especially the new generation of ODS alloys, seem to be promising components for ceramic/metal joints used at high temperatures. Both the Si_3N_4 and the superalloy possess high strength, good creep and excellent corrosion resistance at high temperatures.

Direct bonds of Si_3N_4 to most metals fail during temperature changes /3/ due to the large difference in the thermal expansion coefficients of the components. During cooling from the joining temperature, the contraction of the metal part is usually much larger than that of the ceramic part. Residual stresses develop which may exceed the strength of the ceramic parts themselves, the reaction products or the interfaces of the joints. The classical finite element analysis of Suganuma /3/ revealed those stresses in a Steel/Si_3N_4 joint cooled from temperature after joining. In the central contact area of the metal, tensile stresses build up, whereas compressive stresses evolve in the ceramic component. These stresses suppress crack opening in the ceramic part. However, at the edge of the joint, high tensile stresses were present in the ceramic and similar compressive stresses were estimated for the steel component. In this case, crack formation and detrimental crack growth are likely to start at the surface of the ceramic part. Suganuma proposed to introduce ductile interlayers between the parts to be joined, which deform plastically during joining and cooling allowing stress relaxation near and at the interface.

In bonds of Si_3N_4 to metals, the insufficient chemical adhesion at the metal/Si_3N_4 interface can lead to failure by decohesion of the parts at small stresses. According to the mostly ionic-covalent character of its interatomic bonding, Si_3N_4 possesses very little tendency for electron interaction with materials of strong metallic bonding character. Reactive coatings or thin films on Si_3N_4 are to be used to increase the chemical adhesion at the interface to the adjacent metal. This procedure is often called "metallization".

It seems reasonable for Si_3N_4/metal joints to combine the efforts for metallization and for stress relaxing interlayers by using multilayers and reactive diffusion bonding. In the following, attempts are described to join Si_3N_4 to the ODS superalloy MA 6000 (Inco Alloys International) by diffusion bonding during Hot Isostatic Pressing (HIP). At first, calculations of residual stresses arising during cooling of the Si_3N_4/MA 6000 and Si_3N_4/interlayer/MA 6000 joints are described. Second, adhesion and reactions with various interlayers are reported.

Calculation of residual stresses

A computer program was developed calculating the stress distribution in metal/ceramic strips after cooling. The program is based on the analytical estimation of residual stresses in metal/ceramic strips by Hsueh and Evans /4/. Their calculation method was extended for multilayer joints taking into account the material's behavior of the interlayers. The calculations were used to select combinations of interlayers leading to favourably small residual stress states in the joints at room temperature.

If thin strips of a metal and a ceramic with different thermal expansion coefficient are joined at elevated temperature, the joint will bend during cooling. For the ceramic, it is assumed that only ideal elastic deformation takes place. The stress can be simply related to the strain by

$$\sigma_c = E_c(\epsilon_0 + (\alpha_m - \alpha_c)\ T) + E_c(x - t_n)/r \qquad (1)$$

α_c: thermal expansion coefficient of the ceramic part
α_m: thermal expansion coefficient of the metal part
E_c: Young's modulus of the ceramic part
t_n: position of neutral axis
x : position in the strip
r : radius of curvature

The metal may also deform plastically if the effective stress exceeds the yield stress. Linear work hardening is assumed during plastic deformation. In the case of partial plasticity, the stresses can be calculated for the elastic regions by

$$\sigma_m = E_m\left[\epsilon_0 + (x - t_n)/r\right] \qquad (2)$$

and for the plastic regions by

$$\sigma_m = E_m\left[\epsilon_0 + \frac{\sigma_y}{H} + \frac{x - t_n}{r}\right]\left[1 + \frac{E_m}{H}\right]^{-1} \qquad (3)$$

E_m: Young's modulus of the metal
σ_y: yield stress
H : work hardening rate

The elastic/plastic boundary cannot be calculated analytically, therefore an iterative method is used: in the first step, the radius of curvature of the joint is evaluated for fully elastic behaviour. Secondly, those regions where the effective stresses exceed the yield stress, are treated plastically and a new radius of curvature is calculated.

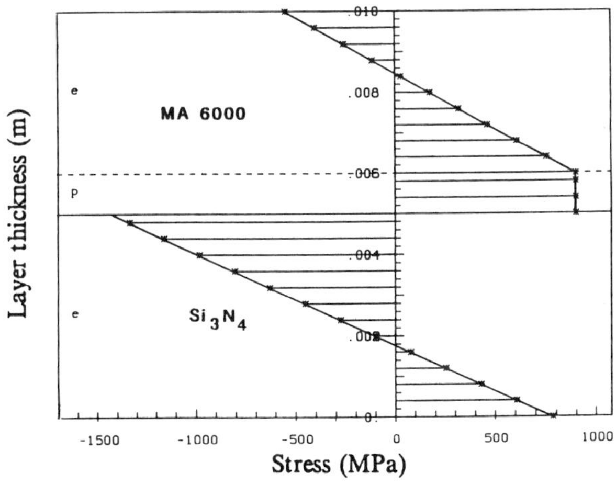

Figure 1 - Stresses in a MA6000/Si_3N_4 joint after cooling from 1200^{0}C.
Yield stress of MA6000, σ_y = 900 MPa.

In the third step, this radius serves as basis for the calculation of a second hypothetical elastic stress distribution implying a change of the elastic/plastic boundaries. This step is repeated until the positions of the elastic/plastic boundaries become fixed.

Figure 1 shows that large tensile stresses arise in the Si_3N_4 directly joined to the superalloy MA 6000 and cooled from 1200^{0}C to room temperature. The thickness of both the ceramic and the metal block was 0.5 cm with a joint interface of 1 x 1 cm^2. At the interface, large tensile stresses of about 1000 MPa are present, which would exceed the strength of the joint leading to crack formation and failure of the joint. At the ceramic surface, tensile stresses are close to 800 MPa.

Assuming a Fe-36wt.%Ni alloy as the joint component with a yield strength of 300 MPa (Figure 2) the tensile stresses at the ceramic surface are reduced to 300 MPa and the compressive stresses at the interface are only of the order of 500 MPa.

If the Fe-36wt.%Ni serves as an interlayer between MA6000 and Si_3N_4, the stresses are somewhat reduced compared to the interlayer free joint, as shown in Figure 3, but it seems fair to state that the stress relaxation effect of those thin interlayers is rather limited as long as elastic-plastic deformation is taken into account and extended creep relaxation is absent.

Experiments

Sheets of Si_3N_4 and MA6000 with thicknesses of 5 mm and the soft metallic interlayers with about 1mm thickness were prepared by cutting from larger bulks. Samples of MA6000/ Fe-36wt.%Ni/Si_3N_4 were produced by diffusion bonding during Hot Isostatic Pressing at 1100 to 1200^{0}C with 50 MPa for 2h. The cooling rate was 10 K/min. For an increased chemical adhesion, some experiments were made with various foils of transition metals, such as nickel, titanium or niobium between the ceramic and the soft metal. Before joining, the surfaces were grinded and polished to an average roughness of 0.1 μm. The MA6000/interlayers/Si_3N_4 combinations were placed into glass capsules which were evacuated and sealed before HIPing in a laboratory HIP equipment. The pressure was applied after the samples had reached the isothermal diffusion bonding temperature. After bonding, the metal/ceramic and metal/metal interfaces were investigated by SEM, EDX and WDX analysis.

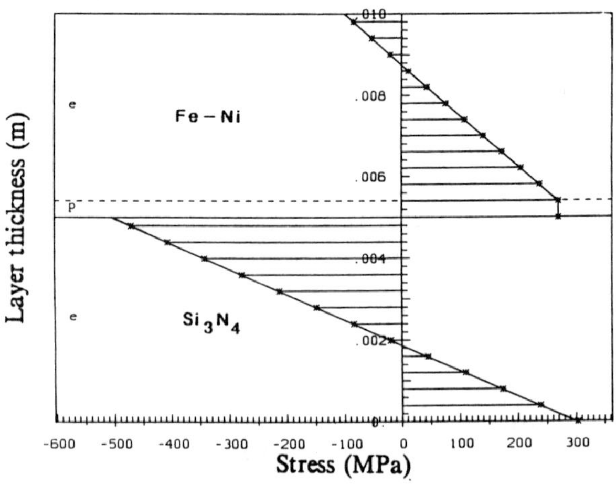

Figure 2 - Stresses in a Fe-36wt.%Ni/Si_3N_4 joint after cooling from 1200^{0}C. Yield stress of Fe-36wt.%Ni, σ_y = 250 MPa.

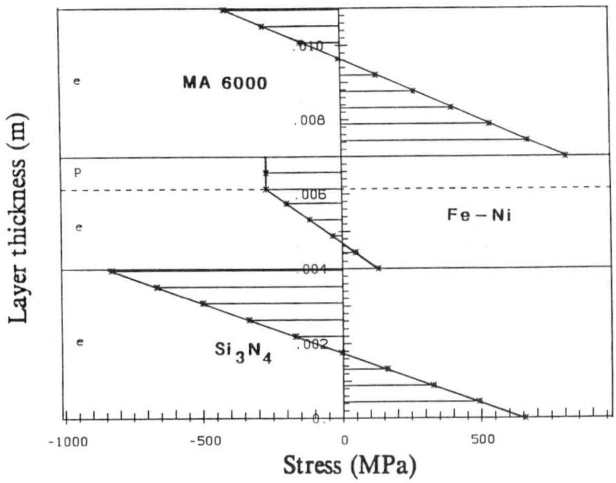

Figure 3 - Stresses in a MA6000/Fe-Ni/Si_3N_4 joint after cooling from 1200°C. Yield stress of MA6000 and Fe-Ni, σ_y = 900 and 250 MPa.

Reactions during joining with interlayers

Figure 4 shows the interface of a Fe-36wt.%Ni/Si_3N_4 joint. The bright (lower) block is Fe-Ni, the upper (dark) block is Si_3N_4. The EDX analysis indicated that Si dissolves to approximately 4 wt.% (appr. 8 at.-%) in the Fe-Ni alloy. The Fe to Ni ratio of 0.52 to 0.56 is similar to the initial alloy. The dissolution of Si in the Fe-36wt.%Ni lead to a large amount of free N_2 (appr. 11 at.% N) because the nitrogen can partly dissolve in the austenite (appr. 0.13 at.%) only.

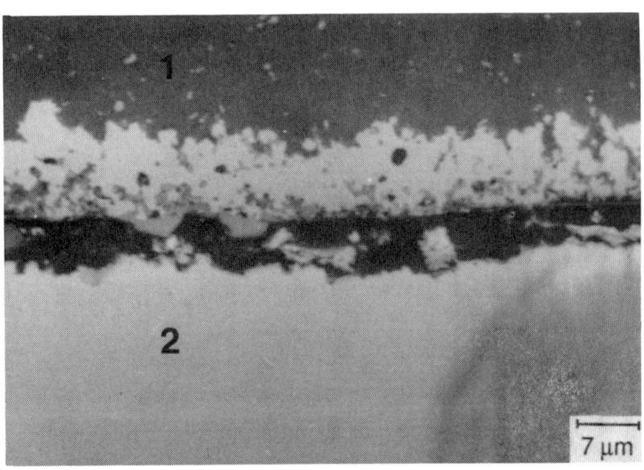

Figure 4 - A transverse cross-section of a Si_3N_4/Fe-Ni interface. 1: Si_3N_4 2: Fe-Ni.

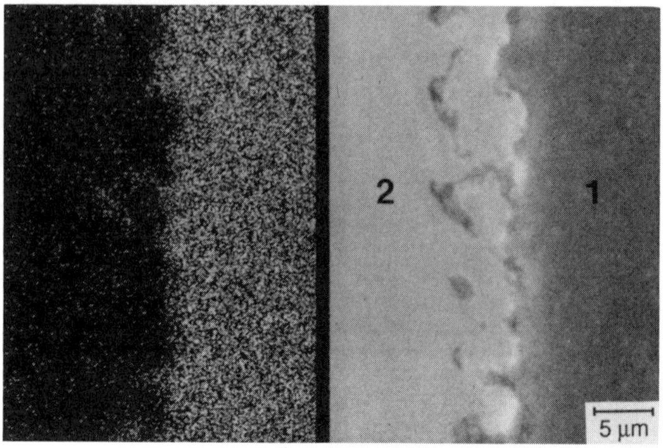

Figure 5 - SEM micrograph and Si distribution at the Si_3N_4/Ni interface.
1: Si_3N_4 2: Ni.

This effect becomes clearly visible in joints of Ni/Si_3N_4, where 6 at.% of Si was dissolved in the Ni and practically no nitrogen is soluble in the Ni matrix. Figure 5 shows a transverse cross-section of a Si_3N_4/Ni interface, HIPed at 1200°C for 1h. The diffusion of Si into Ni produced a Ni(Si) solid solution, but no reaction layer of brittle phases was observed. The high magnification shows that voids formed at the interface. The voids may weaken the interfacial strength of the joint, but it should to be mentioned that the Ni/Si_3N_4 joint appeared to be mechanically quite stable, hence the voids may not be detrimental to the joining area if part of the nitrogen can be trapped by strong nitride formers.

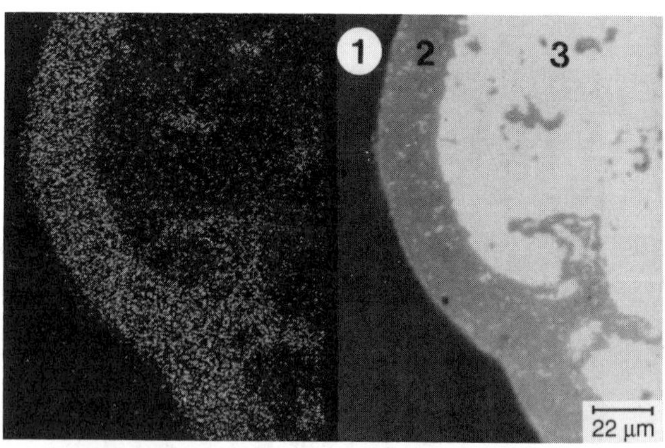

Figure 6 - Ti-distribution at the interface of a Si_3N_4/Ti/Fe-Ni joint.
1: Si_3N_4 2: TiN 3: Fe-Ni.

Thus, joints were produced with reactive metals, such as Ti, V and Nb. Figure 6 shows the interface of Si_3N_4 (right picture, left (darker) part) to the Fe-Ni alloy using titanium (250 μm) as a reactive foil. A reaction layer was observed at the position of the Ti foil of about 10 μm in thickness which proved to be TiN. The Ti content in the other initial areas of the Ti foil was strongly reduced.
Again, Si_3N_4 decomposed, but the free nitrogen reacted with titanium to TiN. No silicides were detected at the ceramic/metal interface. There was also a strong interdiffusion between titanium, iron and nickel at the initial metal/metal interface forming a Ti-Ni-Fe solid solution (Fig. 7). The use of reactive Ti foil seems therefore to be a promising method to increase the chemical adhesion of Si_3N_4 to metals.

However, the formation of TiN led to crack formation with crack propagation along the interface between the two nitrides. The reason for interfacial damage can be found in the difference in the stiffness ($\Delta E = 360$ GPa) and in the thermal expansion ($\Delta\alpha = 6 \cdot 10^{-6}$ 1/K) between Si_3N_4 and TiN. Figure 8 shows the stress distribution of a Si_3N_4/TiN/Fe-36Ni joint calculated with the assumption that during bonding TiN with a thickness of 20 μm formed. The large calculated tensile stresses at the interface indicate that even thin reaction layers of TiN may cause failure of the joint due to the development of high thermal stresses during cooling.

Obviously, the formation of a thin reaction layer with a high Youngs modulus is not favourable for the crack resistance of a joint. On the other hand, the nitrides showed to be effective in forming satisfying bonding layers at the Si_3N_4/metal interface.

A limited activity of the nitride former at the Si_3N_4 interface combined with a low activity in the adjacent layers should allow the formation of extremely thin nitride layers at the interface and suppress the formation of N_2 voids in the adjacent metal bulk. These conditions were realized by applying the interdiffusion metals layer technique proposed by Iino /5/.

A material sequence of Si_3N_4 / Ni / Nb / Fe-36Ni / MA6000 was chosen, where Ni is an Interface Metal and Nb is the Adjacent Metal representing the nitride former. The thicknesses of the Ni foil and the Nb sheet were 10μm and 1mm, respectively. Figure 9 shows a transverse cross-section of the metal/ceramic interface after joining. During bonding, the Adjacent Metal Nb diffuses into the Interface Metal forming an extremely thin Nb-nitride layer at the Si_3N_4 interface. The interdiffusion of Nb and Ni resulted in the formation of a number of intermetallic phases. EDX analysis indicated a sequence of Ni_8Nb - Ni_3Nb - (Ni_3Nb + Ni_6Nb_7) - Ni_6Nb_7 - (Nb-Fe-Ni).

The mechanical stability of the joint was excellent despite the fact that fine cracks were visible in the region of the intermetallic phases. These cracks run perpendicular to the metal/ceramic interfaces. In contrast, other interlayer systems had shown cracks parallel to the interface of the joint which obviously led to a high susceptibility to catastrophic interfacial damage.

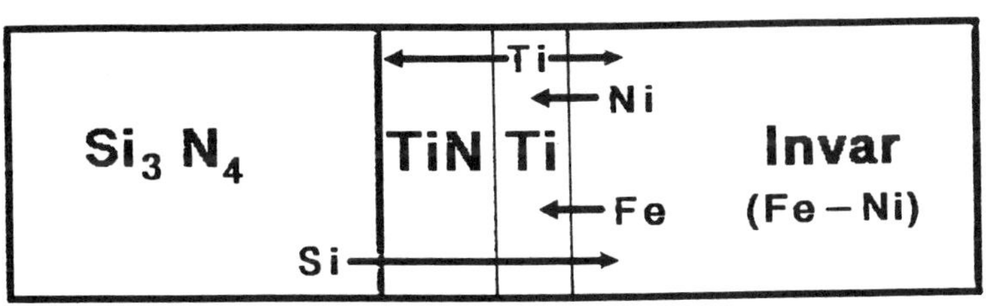

Figure 7 - Interdiffusion at the Si_3N_4/Ti/Fe-Ni interface (schematically).

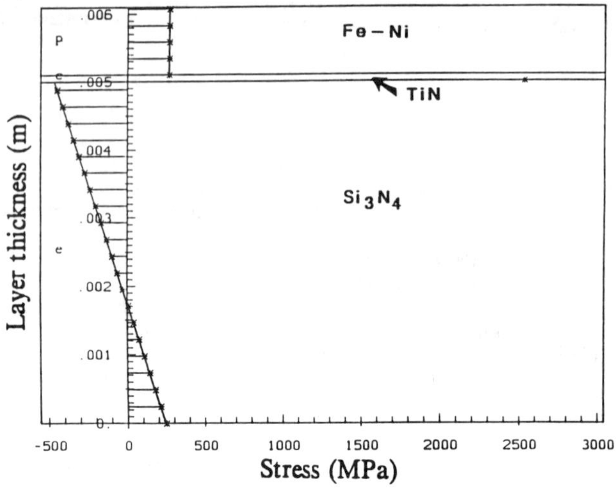

Figure 8 – Stress distribution in a Si_3N_4/TiN/Fe–Ni joint.

It is thought that the fine perpendicular cracks may cause a partial relaxation of thermal stresses in the joint. They appear to be beneficial in suppressing the interfacial damage along the metal/ceramic interface.

Another example is shown in Figure 10 in which perpendicular microcracks were found beneficial for the mechanical stability of the joint. It shows the microstructure of an initial layer sequence Si_3N_4/Hf/ Fe–36Ni/MA6000 in the joint area. In this case, a Hf–silicide layer was formed at the Si_3N_4 interface during bonding leading to a Si_3N_4/$HfSi_2$/Si–Hf–Fe–Ni/MA6000 sequence. Although the presence of the brittle $HfSi_2$ layer (thickness appr. 1 μm) would suggest to provoke the decohesion of the

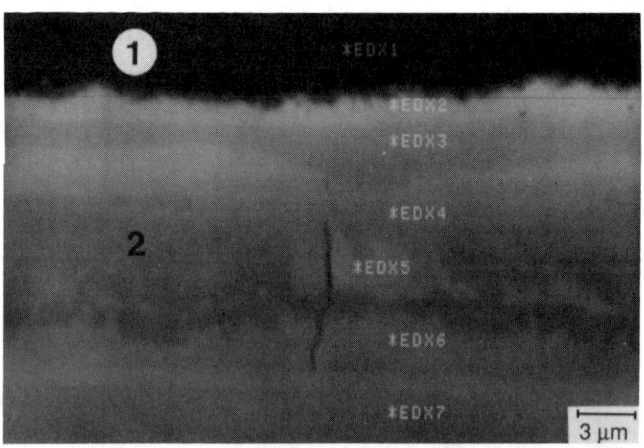

Figure 9 – A transverse cross–section of the Si_3N_4/Ni/Nb/ interface. The markers indicate the positions for EDX analysis. 1: Si_3N_4 2: initial Nb–foil.

diffusion bonded joint, the stresses are obviously relaxed by small cracks which may lead to an island structure in the brittle phase. Such fine microcracks were also observed in the reaction zones at metal/metal interfaces. It is concluded that in systems with a low ability for stress reduction by creep or plastic deformation, stress relaxation induced by microcracks in stiff reaction layers could be a concept to prevent interfacial damage.

Figure 11 shows schematically two ways to reduce tensile stresses arising in multilayered joints with a stiff interlayer ($E_2>E_1$) during thermal or mechanical treatment. If the stiff reaction layer possesses a high mechanical strength, the stress values at the interface may exceed the interfacial strength and result in stress relaxation along the interface by crack formation between the constituents. Another possibility is that the interfacial damage may be prevented by non-critical crack formation, if the stresses exceed the mechanical strength of the stiff reaction product. At first, microcracks may form perpendicular to the interface. Secondly, an additional stress relaxation could be obtained by crack blunting. Which of the mechanisms will proceed is influenced by the ratio of the Youngs modulus, the thickness of the reaction product and the ratio of the interfacial strength to the strength of the interlayer.

Conclusions

Si_3N_4 and the ODS-superalloy MA6000 are potential components for metal/ceramic joints at elevated temperatures. Due to the large thermal mismatch between both components, joints with interlayers are required which allow the relaxation of thermal stresses arising during thermal treatment. Calculations of the stress development and experimental investigations have shown that stress relaxation by thin soft interlayers is very limited in Si_3N_4/MA6000 joints and also in similar systems. During bonding of Si_3N_4 to metals, which are suitable as interlayers, brittle reaction products frequently formed at the metal/ceramic interface. The thickness of these layers could be controlled by an Interdiffusion Layer Technique which limited the activity of the reactive metals.

In systems with a low deformability of the stiff reaction layers, stress relaxation by controlled microcrack formation in these layers appears to reduce the interfacial damage and to improve the mechanical stability of the joints.

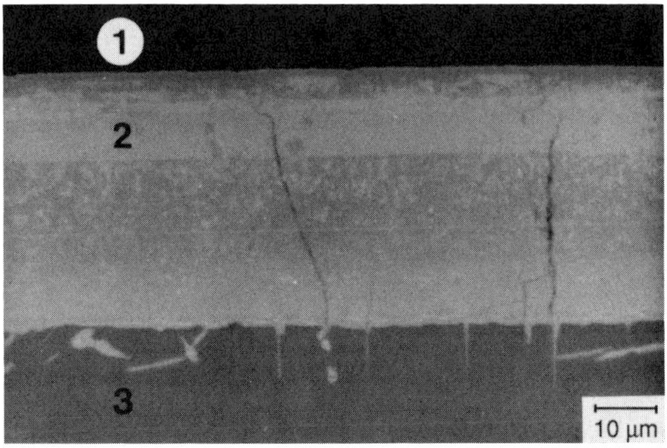

Figure 10- SEM micrograph for a Si_3N_4/Hf/Fe-Ni interface.
1: Si_3N_4 2: initial Hf-foil 3: Fe-36wt.%Ni.

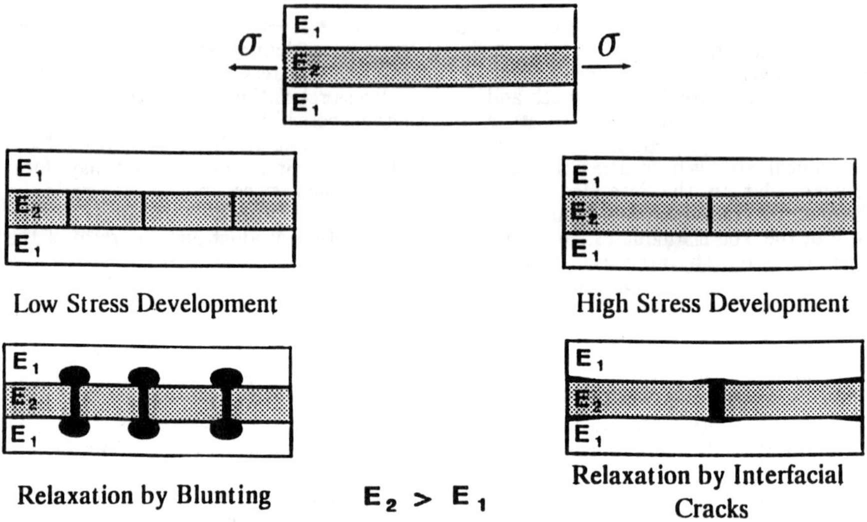

Figure 11 - Two possibilities for stress relaxation in systems with stiff reaction layers.

Acknowledgement

This research was supported by the Volkswagen-Stiftung and by G.I.F., the German-Israeli Foundation for Scientific Research and Development.

References

1. W.E.Borbigde, R.V.Allen and P.T.Whelan,"A Review of the Bonding Technique for Joining Ceramics to Metals", J. de Physique, 47 (1986), C1-131-137.

2. M.G.Nicholas and D.A.Mortimer,"Ceramic/Metal Joining of Structural Applications", Mater.Sci.and Tech., 1 (1985), 657-665.

3. K.Suganuma, T.Okamoto, and M.Koizumi,"Effect of Interlayers in Ceramic-Metal Joints with Thermal Expansion Mismatches", J.Am.Ceram.Soc., 67 (1984), C256-C257.

4. C.H.Hsueh and A.G.Evans,"Residual Stresses in Metal/Ceramic Bonded Strips", J.Am.Ceram.Soc., 68 (1985), 241-248.

5. Y.Iino and M.Taguchi, "Interdiffusing Metals Layer Technique of Ceramic Metal Bonding", J.Mat.Science. Letters, 7 (1988), 981-982.

RESIDUAL STRESSES IN METAL/CERAMIC BRAZES:

EFFECT OF CREEP ON FINITE ELEMENT ANALYSIS RESULTS

J. J. Stephens, S. N. Burchett and F. M. Hosking

Sandia National Laboratories
Albuquerque, NM 87185-5800

Abstract

Creep-induced stress relaxation of a braze alloy can be an important factor in successful processing of certain metal/ceramic braze joint geometries where there is a mismatch in thermal expansion. The inclusion of a creep constitutive law in finite element analysis (FEA) codes makes it possible to calculate the effect of a particular cooling rate from the braze temperature on residual stresses in the ceramic. Valid calculations require a constitutive law for creep of the brazement which accurately predicts the minimum strain rate over a wide range of temperatures. Using data from previous investigations on the creep of recrystallized Cu, we have developed minimum strain rate creep correlations at both high and intermediate temperatures for pure Cu. At the higher temperatures, a power law is used, with the activation energy obtained by regression analysis quite close to the value of lattice self diffusion in Cu. At intermediate temperatures (350-600°C), we have found that the Garofalo sinh equation obtains the best fit to the data, with an activation energy approximately equal to 2/3 of the activation energy for lattice diffusion, attributable to dislocation core diffusion. Results for FEA calculations of the residual stresses obtained in Cu brazing of a metallic feedthrough pin to a metallized alumina ceramic are presented using both time-independent and creep constitutive laws for pure Cu. At higher temperatures, the creep law predicts lower residual stresses in the alumina compared to the elastic/plastic constitutive law. However, at lower temperatures, the results obtained with the elastic/plastic constitutive law indicate lower stress levels in the ceramic compared to the predictions obtained with the various creep constitutive relations. The lower temperature results are discussed in the context of primary creep behavior.

This work conducted at Sandia National Laboratories, supported by the U. S. Department of Energy under contract number DE-AC04-76DP00789.

Note: "Kovar" is a registered trademark of Carpenter Technology Corporation, and "Palco" is a trademark of Wesgo Division, GTE Products Corporation.

Metal - Ceramic Joining
Edited by P. Kumar and V. A. Greenhut
The Minerals, Metals & Materials Society, 1991

Introduction

Residual stresses resulting from metal/ceramic brazing are of interest for a number of vacuum electronic component applications because high tensile stresses in the ceramic may result in cracking and loss of hermeticity. The residual stresses are caused by the use of materials with different coefficiencts of thermal expansion, along with relatively high braze alloy melting points. For example, the vacuum seals which we will discuss in this paper use electronic grade Cu as the brazement, which has a melting point of 1084°C. For the case of soldered electronic assemblies incorporating alumina substrates, previous stress predictions based on Finite Element Analysis (FEA) at our laboratory (1,2) have demonstrated that the residual stress in the alumina ceramic can be minimized by proper control of cooldown from the soldering temperature if a minimum creep rate equation for the solder is included as part of the materials parameters used for the calculations. A similar effect is expected for the case of metal/ceramic braze joints, with the important qualification that the amount of creep-induced stress relaxation which can be expected depends on braze joint geometry as well as the temperature cooldown profile. The need for incorporation of creep induced stress relaxation into FEA calculations for braze joints has also been identified by Charreyron, et. al. (3) for the case of the Ag-Cu eutectic alloy. Charreyron, et. al. also pointed out that creep data for eutectic Ag-Cu alloy are not available. Such data have been recently generated at Sandia National Laboratories (SNL), and will be available as part of a forthcoming publication (4).

In general, a large number of variables will control the level of residual stresses realized in a metal/ ceramic brazing process, including materials selection, thermal expansion coefficients, the amount of braze and the brazing time, temperature and cooldown profile. However, in addition to these basic considerations, the amount of hydrostatic constraint present in the joint design is extremely important. In general, those braze joint geometries with a large deviatoric stress component and relatively minor hydrostatic component have the greatest potential for stress relaxation during the cooldown portion of the braze cycle. This is illustrated in Figure 1, where we have drawn in schematic fashion three classes of geometry often encountered in metal/ceramic braze joints with cylindrical symmetry. Those braze joints of the "Shear" variety shown in Figure 1a generally have the greatest opportunity for stress relaxation because of small hydrostatic stresses and significant deviatoric (ie., shear type) stresses. The second class of braze joints, those termed "Tension- Compression" and illustrated in Figure 1b, have more limited capacity for stress relaxation relative to joints with the "Shear" geometry. Finally, the third class of braze joints shown in Figure 1c, termed "Tortuous Path" joints, tend to have relatively high hydrostatic stresses - especially at the 90° corners - and consequently have very little potential for stress relaxation. It should also be pointed out that those geometries with greater oppotunity for creep induced stress relaxation are of a somewhat complicated geometry, which makes experimental verification of code results difficult without subjecting parts to metallographic cross sectioning or other destructive inspection for cracks. While a simple slab geometry has been used for X-ray residual stress analysis measurements in braze joints (5,6), such experiments are undoubtedly influenced by geometry as well as by materials selecton and processing. In addition, due to the relatively low penetration of X-rays into the bulk of a specimen, it is necessary to grind successive sections into the sample in order to obtain detailed information with X-ray residual stress analysis.

In the current paper, we will be presenting the methodology which has been developed for the calculation of residual stresses in metal/ceramic braze joints corresponding to the "Shear" type geometry described above. In the braze joint geometry of specific interest to us, shown in cross section in Figure 2a, we are concerned with area seals which use an alumina ceramic that has been metallized using Mo/Mn metallization, along with a subsequent Ni plate prior to brazing. The parts are brazed with electronic grade Cu, a Kovar®™ washer in the middle of the brazement to reduce the amount of thermal strain, and a pin feedthrough wire which has been made of Kovar as well as other alloys which are readily wet by Cu. We have observed two specific types of cracking in this type of area seal braze joint, as shown in Figure 2b and 2c. In both cases the cracks in the alumina ceramic originate near the very end of the Cu brazement, which is where the maximum residual stresses in the ceramic are encountered. "Type A" cracks propagate inward into the bulk of the ceramic at an angle of about 30-45° from the metallization interface and are most commonly observed when high residual stresses are present. A second,

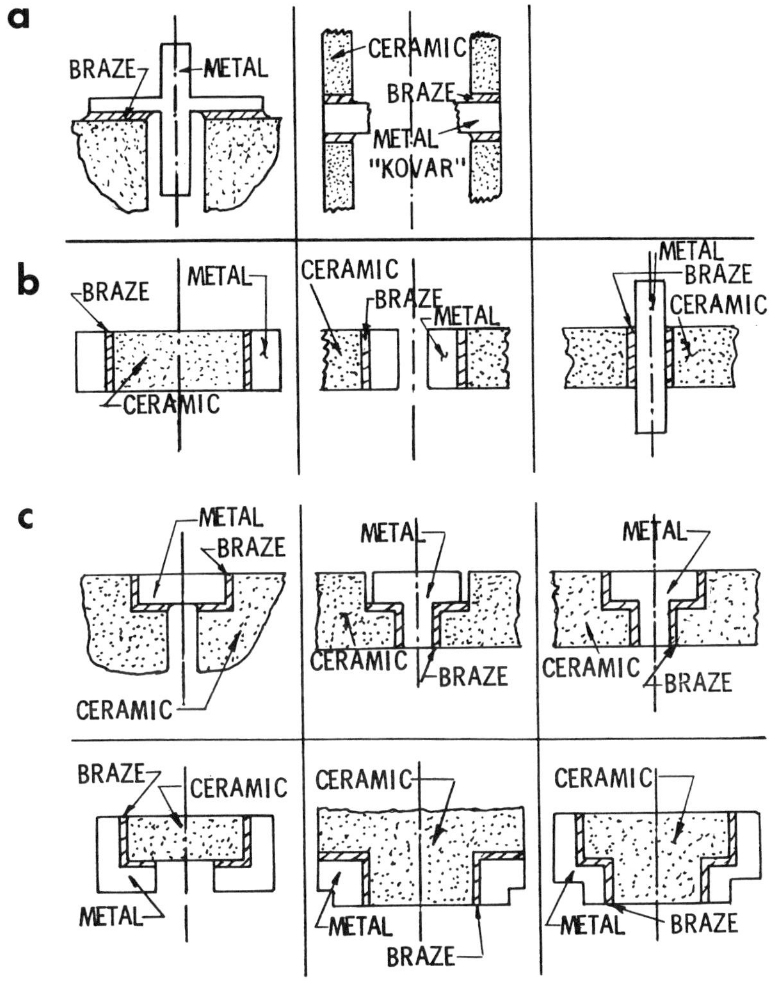

Figure 1 - Generic metal/ceramic braze joints, grouped according to their potential for creep induced stress relaxation during cooldown from the brazing cycle. All sketches depict components with cylindrical symmetry, with the axis of symmetry (z-axis) indicated. (a) Two types of "Shear" braze joints, where the dominant loading is in shear, with good potential for stress relaxation. (b) Three types of "Tension/Compression" braze joints with more limited potential for stress relaxation relative to (a). (c) Six types of "Tortuous Path" braze joints, where there is a significant hydrostatic stress component and correspondingly little potential for stress relaxation.

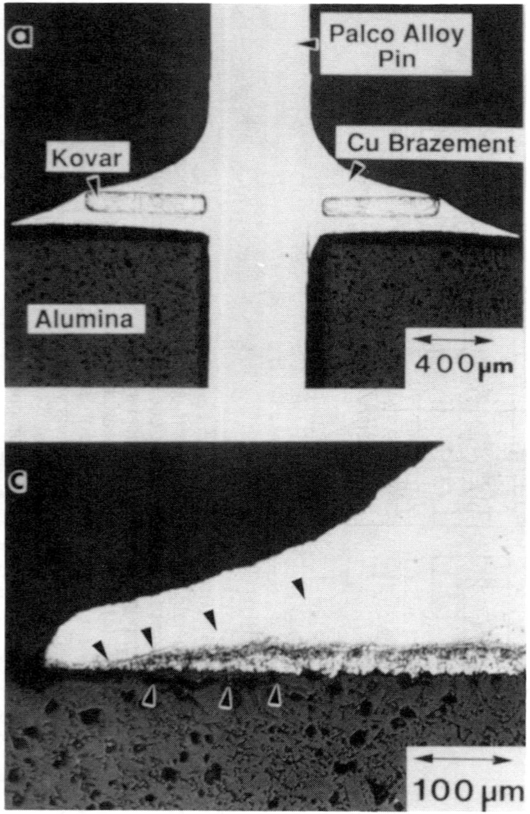

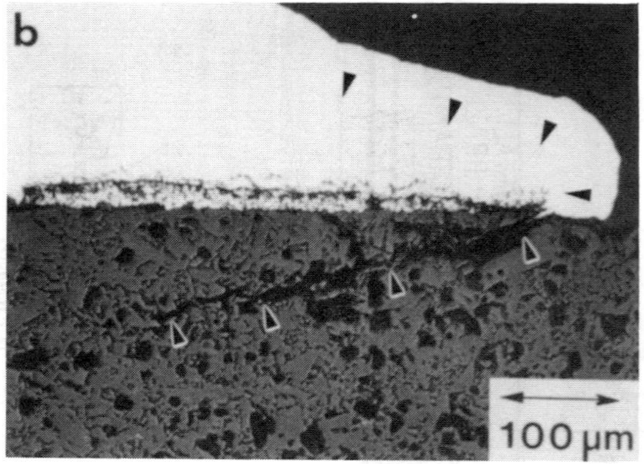

Figure 2 - Metallographic cross sections (diametral plane) of the "area seal" discussed in this paper. (a) Overall view, with the materials of construction indicated. This area seal was brazed in wet hydrogen for a total of 3 minutes above 1084°C, with a peak temperature of 1097°C and a furnace cooldown shown in the right hand side of Figure 3. Both (b) and (c) are from different samples and have a layer of Ni plated onto the Cu brazement - the Cu/Ni interface is marked by the upper set of arrows. (b) "Type A" crack (lower arrows) proceeding into the bulk of the alumina ceramic from the end of the brazement. (c) "Type B" crack (lower arrows) running along the interface between the Mo-Mn metallization and the alumina ceramic.

less commonly observed crack morphology has been termed "Type B" which takes a path along the metallization/ceramic interface The former type of cracking is considered more problematic as considerably longer crack lengths are observed with the Type A as opposed to the Type B cracks.

For the electronic components of interest, existing production equipment dictates these parts be brazed using wet hydrogen belt furnaces. A typical belt furnace temperature profile of the brazing cycle is shown in Figure 3. Because of the significant time spent above the melting point of Cu (1084°C), belt furnace processing can lead to significant amounts of alloying of the Cu brazement; with the exact level depending on the pin material selection. We have observed generally lesser amounts of alloying of the brazement with Kovar pins: this is related to the amount of solubility which is possible in the Cu brazement for a given pin material. For example, Ni-plated 316L stainless steel or Palco™ (65Pd-35Co) alloy pins generally lead to higher levels of alloying in the brazement than with Kovar alloy pins. The results of an extensive electron microprobe study of the effect of braze time, temperature, and pin material on alloying levels in the Cu braze are documented in a separate publication (7). In particular, batch furnace brazing of these area seals with minimal time spent above 1084°C can reduce the average level of alloying in the brazement to approximately 1-2 wt.%, even with the Palco alloy which leads to considerable alloying of the Cu as a result of the conditions typical of the belt furnace braze process. Figure 4 illustrates the effect of increasing levels of Ni in Cu on the stress required to maintain identical creep rates at 727°C (8,9). While the alloying of Cu undoubtedly plays an important role in the residual stresses observed in the ceramic, the fact that creep data for the alloyed braze compositions which we have observed are not available precludes any analytical treatment of alloyed brazes in the present paper. A separate experimental program which is investigating the creep behavior of Cu solid solution alloys which are typically obtained as a result of specific brazing conditions is currently in progress (10).

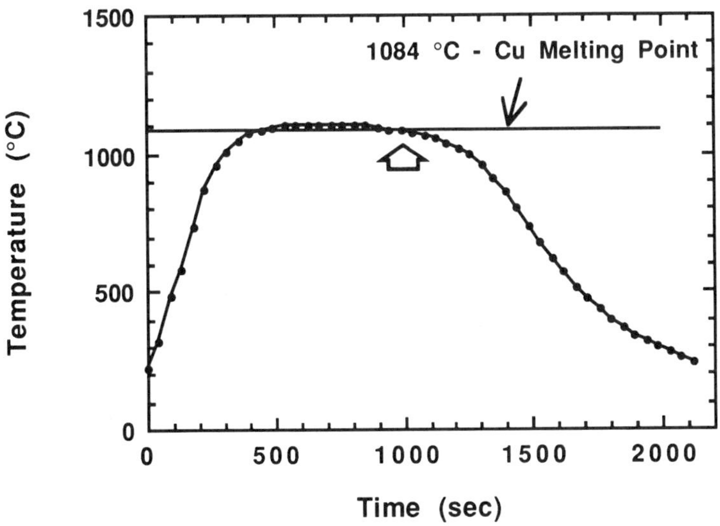

Figure 3 - Typical belt furnace profile used for Cu brazing of area seals. This particular profile has a peak temperature of 1109°C with a total of about 10 minutes above 1084°C. The cooldown portion of this figure is used as the baseline "cooldown" cycle for the FEA calculations presented in this paper. Extrapolation to room temperature yielded a total cooldown time of 1700 seconds.

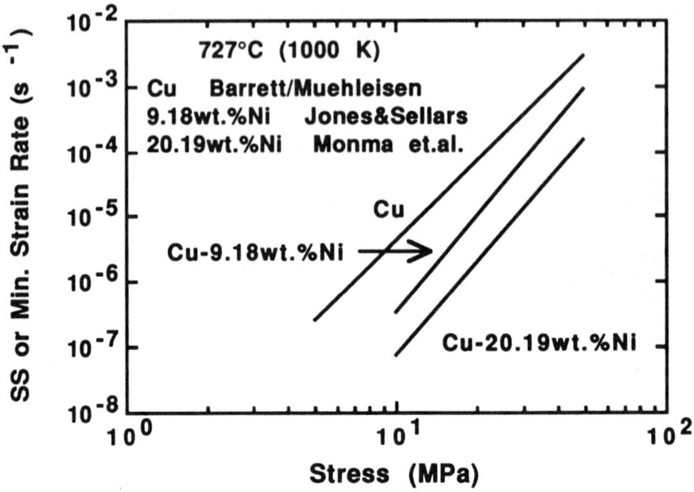

Figure 4 - The effect of Ni on the predicted minimum creep rate behavior of Cu at 727°C (1000 K). Data for the 9.18 wt.% Ni alloy are from Jones and Sellars (8) and for the 20.19 wt.% Ni alloy are from Monma, et. al (9). The correlation for unalloyed Cu is based on eq. (4).

Scope of the Present Study

In this paper we will concentrate primarily on development of two related topics: 1. adequate minium creep rate constitutive relations for creep of pure Cu and 2. the computational methodology required to address the generic residual stress problem. In addition to using the creep correlations for pure Cu to compute residual stresses, we will also compare those results with the results for calculations which treat the Cu as either an elastic or elastic/plastic material. We will examine two different geometries which represent extremes in the amount of Cu brazement which could be realized in actual braze joints.

Minimum Creep Rate Correlations for Pure Cu

General Considerations

Uniaxial creep data from three separate investigations (11-13) were used, along with multivariable linear regression analysis, to construct minimum creep rate correlations for pure Cu. The data in these three references spans the temperature range from 350-1058°C (homologous temperature (T/T_m) of 0.46-0.98) and a stress range of 0.23-100 MPa. In an attempt to achieve consistency with the brazing process, we have limited the data used in these correlations to recrystallized Cu samples with a grain size of $\geq$ 100 μm, as the grain size of lightly alloyed Cu braze joints tends to be fairly large. Barrett, et. al. (14) examined the effect of grain size on the tensile creep behavior of Cu at 406, 496 and 626°C, and found that the minimum or steady state creep rate is essentially independent of grain size for annealed samples with grain sizes in excess of 100 μm.

We have found that in order to be able to adequately fit the minimum creep rate data for Cu over such a wide range of stress and temperature, it is necessary to separate the data into high and intermediate temperature regimes. Data obtained at temperatures greater than approximately 600°C (T/T_m=0.64) are considered "high temperature" data, while data obtained up to 600°C are considered "intermediate temperature" data. This demarkation point was arrived at by means of trial and error, but is consistent with the observations of Barrett and Sherby (11) with respect to the temperature dependence of the true activation energy for creep, Q_c. Barrett and Sherby observed Q_c for Cu to increase to a value approximately equal to the activation for lattice self-diffusion at temperatures in excess of 600°C, along with lower values of Q_c below 600°C.

In the paragraphs that follow, we will discuss the minimum strain rate correlations developed for each of the temperature regimes. The FEA codes use constitutive relations for creep deformation where the applied stresses are not normalized - for example, the power law creep equation for the (uniaxial) minimum strain rate $\dot{\varepsilon}_{min}$ is written as

$$\dot{\varepsilon}_{min}\ (s^{-1}) = A\ (\sigma)^n \exp(-Q_a/RT) \tag{1}$$

where σ is the (uniaxial) applied stress in MPa, T is the absolute temperature in Kelvin, R is the gas constant (1.987 cal/mole/K), n is the stress exponent, Q_a (cal/mole) is the "apparent" activation energy for creep, and A ($MPa^{-n}\ s^{-1}$) is a constant. The values of A, n and Q_a are obtained directly from multivariable linear regression analysis. To facilitate comparison of the activation energies for creep obtained for the various correlations with fundamental diffusion activation energies for pure Cu, we will also present the results of minimum strain rate correlations obtained where the uniaxial applied stress is normalized by the value of the temperature dependent Young's Modulus for Cu:

$$\dot{\varepsilon}_{min}\ (s^{-1}) = A'\ (\sigma/E)^{n'} \exp(-Q_c/RT) \tag{2}$$

where Q_c is the true activation energy for creep, and the constants A' (s^{-1}) and n' will also have values which are different than those of eq.(1) when examining the same data set. This "temperature normalization" technique for stress has been shown by Sherby and co-workers (15) to be the preferred method of obtaining Q_c. For the stress-normalized correlations, we have used the following expression for the dynamic Young's Modulus data for Cu:

$$E\ (MPa) = 1.251 \times 10^5 - 18.58{*}T(°C) - 0.0872{*}T^2 + 4.368 \times 10^{-5}{*}T^3 \tag{3}$$

based on the data obtained by Barrett and Sherby (11).

<u>High Temperature Power Law Equation</u>

The compression creep data of Muehleisen, et. al. (12) which span the temperature range 1024-1058°C and the coarse grained (≥ 100 μm) tensile creep data of Barrett and Sherby (11) in the temperature range of 589 to 950°C were used to develop the following correlations for high temperature creep of Cu:

$$\dot{\varepsilon}_{min}\ (s^{-1}) = 202.1\ (\sigma)^{4.067} \exp(-53{,}814/RT) \tag{4}$$

$$\dot{\varepsilon}_{min}\ (s^{-1}) = 1.073 \times 10^{21}\ (\sigma/E)^{4.172} \exp(-46{,}241/RT) \tag{5}$$

Correlation coefficients (r^2) of 0.972 and 0.978 were obtained for eqs. (4) and (5), respectively. It should be noted that the activation energy for self diffusion in Cu over the temperature range 685-1060°C is 47.12 kcal/mole (16); the value of 46.24 kcal/mole obtained for Q_c in eq. (5) is quite close to this value. By rearranging eq. (4) so that the Arrhenius factor is on the left hand side of the equation - obtaining the Zener-Hollomon parameter (15) - we can compare the data set and fit to the data using one trend line on a log-log graph. This comparison is shown in Figure 5, where it is apparent that there is an excellent correspondence between the data and eq. (4). The stress exponents obtained for eqs. (4) and (5) - approximately 4.1 and 4.2, respectively - are lower than the stress exponent of 4.4 obtained by Barrett and Sherby at 886°C (see Figure 4 of ref. 11); however, the majority of their data consisted of tests run at a selected number of stress levels, at various temperatures, and are more amenable to plotting directly on an Arrhenius plot.

<u>Intermediate Temperature Power Law Equation</u>

The tensile creep data of Raj and Langdon (13) from 350 to 600°C, along with the limited creep data for coarse grained Cu at temperatures <600°C generated by Barrett and Sherby (11), were used to develop correlations for creep of Cu at "intermediate" temperatures. The following

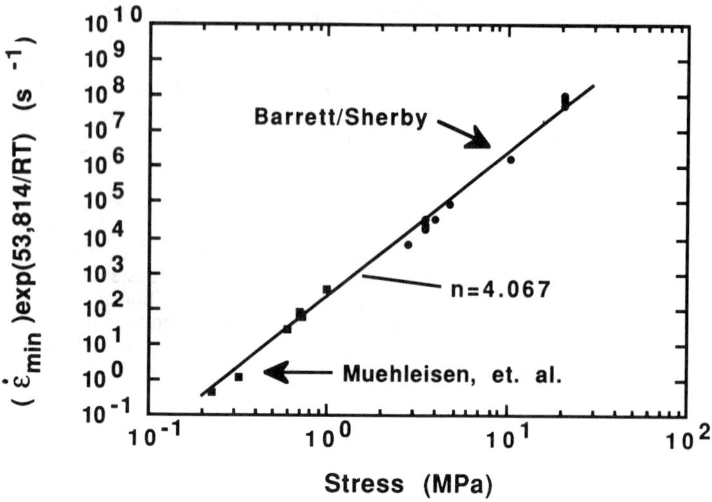

Figure 5 - Zener-Hollomon plot of the high temperature creep correlation for coarse grained Cu shown in eq. (4). The data from Barrett and Sherby (11) include only those data points included in the correlation ($T \geq 589°C$).

power law equations were obtained using multivariable linear regression analysis:

$$\dot{\varepsilon}_{min}\ (s^{-1}) = 1.115 \times 10^{-2}\ (\sigma)^{4.245} \exp(-35{,}402/RT) \quad (6)$$

$$\dot{\varepsilon}_{min}\ (s^{-1}) = 2.456 \times 10^{18}\ (\sigma/E)^{4.269} \exp(-32{,}074/RT) \quad (7)$$

Correlation coefficients of 0.945 and 0.949 were obtained for eqs. (6) and (7), respectively. It is interesting to note that the stress exponent for Cu in this regime is very close to that of the high temperature regime. This behavior is different from Ag, where there is a pronounced change in stress exponent between the high and intermediate temperature regimes (17). The value of Q_c for the intermediate temperature regime (32.07 kcal/mole) is very close to 2/3 of the activation energy for lattice self diffusion in Cu. This is consistent with previous observations for Cu (11), and is generally attributed to the diffusion of vacancies along dislocation cores.

The correlation between eq. (6) and the creep data of Raj and Langdon (13) is shown in Figure 6. We have also included the data and predictions based on eq. (6) for the data at 650 and 700°C, where it is apparent that the intermediate temperature regime correlation may be extrapolated reasonably well up to 650°C, but the prediction systematically underestimates the minimum creep rate at 700°C. This result is synonomous to saying that a higher value of Q_a would be required to adequately fit the data at 700°C. Another important observation from Figure 6 is that the correlation does not fit the data very well at the highest stresses for the 350-500°C data, apparently where the onset of "power-law breakdown" (15) occurs.

Intermediate Temperature Garofalo Sinh Equation

In order to better fit the entire intermediate temperature regime data for Cu, we examined the use of a different power law equation - developed in detail by Garofalo (18) - which uses a hyperbolic sine function:

$$\dot{\varepsilon}_{min}\ (s^{-1}) = A''\ (\sinh(\alpha\sigma))^{n''} \exp(-Q'_a/RT) \quad (8)$$

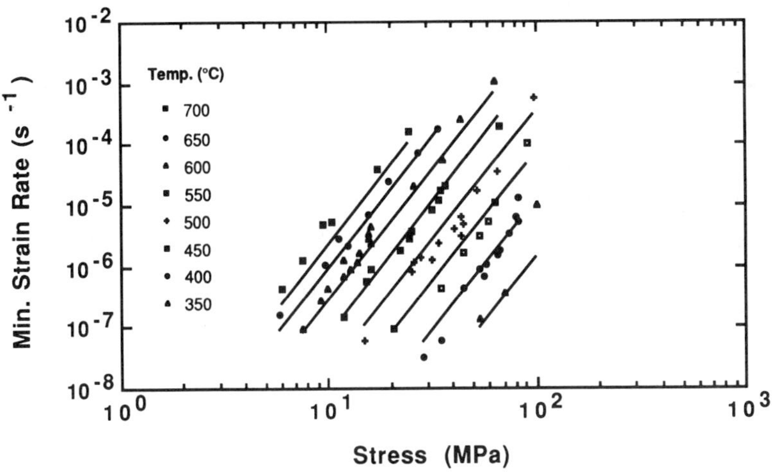

Figure 6 - Comparison of the power law correlation (eq. 6) obtained from Cu creep data at intermediate temperatures (350-600°C) with the entire data set generated by Raj and Langdon (13). Note that the correlation works well for 650°C data, but clearly underestimates the creep rate at 700°C. At lower temperatures (350-500°C) and high stresses, the correlation given by eq. (6) does not work well due to the onset of "power law breakdown."

where A" (s^{-1}), n" and Q'_a (cal/mole) are constants obtained by multivariable regression analysis, and α is a parameter with units of $(MPa)^{-1}$ which must be iterated in order to obtain the highest correlation coefficient. As in the case of the power law creep equation eq. (1), eq.(8) can also be re-written with the stress normalized by Young's modulus. Using the same data set as for eqs. (6) and (7), we obtained the following results using the Garofalo sinh equation:

$$\dot{\varepsilon}_{min}\,(s^{-1}) = 4.805 \times 10^4\,(\sinh(0.0240^*\sigma))^{3.380}\,\exp(-35{,}869/RT) \quad (9)$$

$$\dot{\varepsilon}_{min}\,(s^{-1}) = 3.351 \times 10^3\,(\sinh(2460.^*\sigma/E))^{3.345}\,\exp(-31{,}991/RT) \quad (10)$$

Correlation coefficients (r^2) equal to 0.971 and 0.970 were obtained for eqs. (9) and (10), respectively. It should be noted that the value of Q_c for the sinh law correlation (31.99 kcal/mole) does not differ significantly from that of the power law creep equation (32.07 kcal/mole), however, the quality of fit at higher stresses and lower temperatures is significantly better. This is illustrated in Figure 7a, where the creep data of Raj and Langdon from 350-600°C are compared to the predictions based on eq. (9). In comparing Figure 7a with Figure 6, note especially the better correlation obtained with the sinh law at high stresses in the temperature range 350-600°C. All of the data from (11) and (13) in the temperature range 350-600°C are plotted in Figure 7b using the Zener-Hollomon parameter version of eq. (10) in order to demonstrate the correlation for the case where the stress is normalized by the elastic modulus.

Finite Element Analysis Calculations Incorporating Creep Constitutive Laws

General Remarks

The creep constitutive relations developed above for coarse grained Cu were used to incorporate explicit time dependence into the calculation of residual stresses in alumina ceramic as a result of specific cooldown histories. The exact combinations of constitutive relations used for the calculations are discussed in this section. As indicated above, the driving force for the

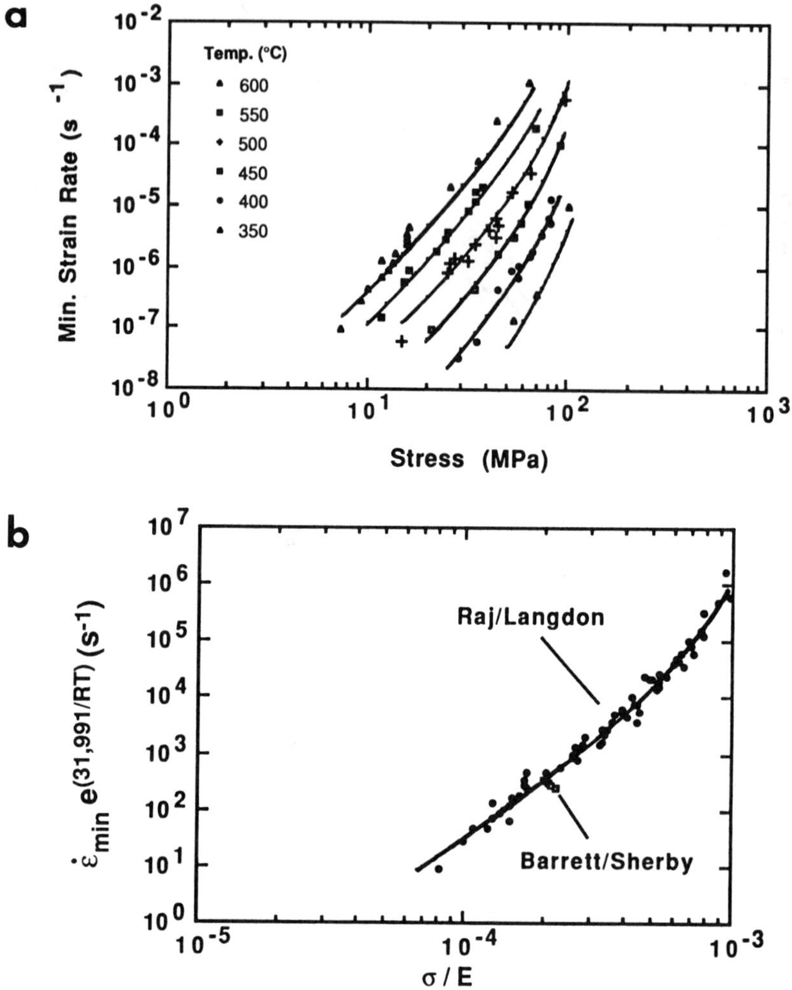

Figure 7 - (a) Comparison of the fit obtained with the Garofalo sinh law (eq.(9) and the 350-600°C creep data of Raj and Langdon (13). (b) Zener-Hollomon plot of eq. (10) in comparision with the entire 350-600°C data set (11, 13) used for the correlation. Note that the value of Q_c (31.99 kcal/mole) is very close to 2/3 of the activation energy for self diffusion in Cu (47.12 kcal/mole).

development of residual stresses in the ceramic is the difference in thermal expansion of the various materials used in the area seal braze joint; the thermal strains are plotted on heating from room temperature for Cu, Kovar and alumina are shown in Figure 8a. Note that the difference in thermal strain between Kovar and alumina is fairly small up to ~650°C, but above this temperature the coefficient of thermal expansion for Kovar increases substantially (19). Because of its relatively high coefficient of thermal expansion at all temperatures, the thermal strains observed for Cu are much higher than for the other two materials in Figure 8a. In addition, the elastic modulii for these three materials are also quite different, as shown for the case of Young's Modulus in Figure 8b (20, 21).

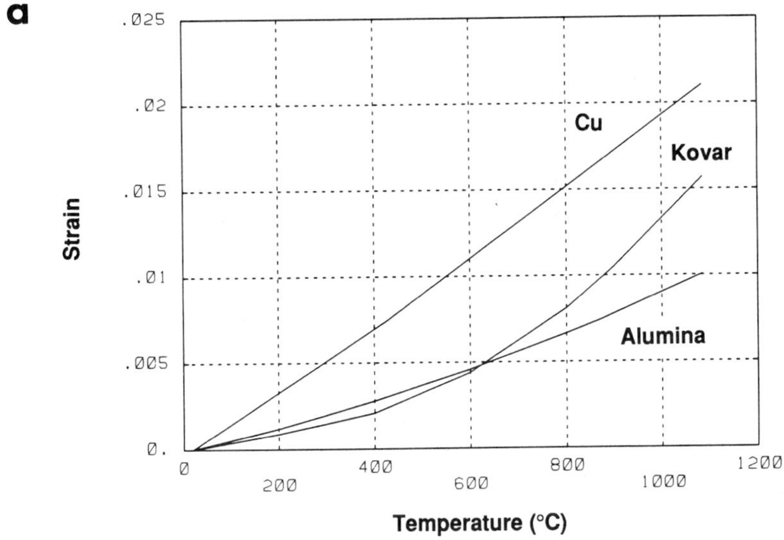

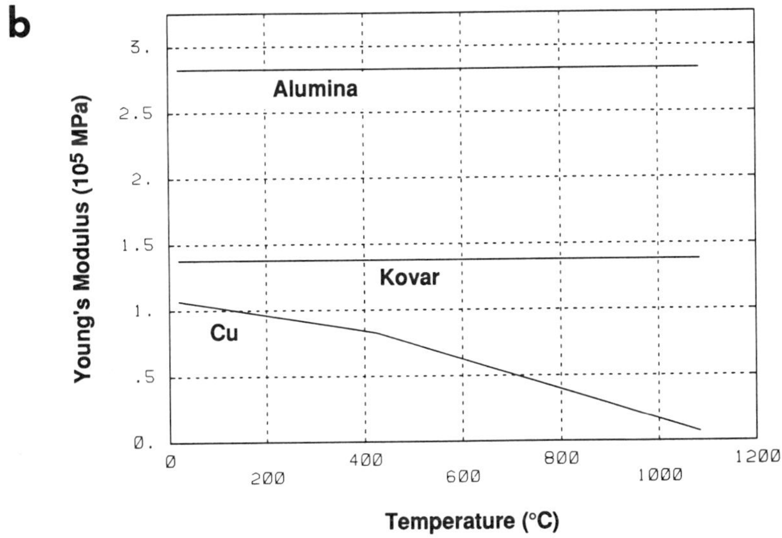

Figure 8 - (a) Thermal strain (on heating) for Cu, Kovar alloy and alumina from room temperature up to the melting point of Cu (1084°C). (b) Values of dynamic Young's modulus for Cu, Kovar alloy and alumina which were used for the FEA calculations.

All FEA results were obtained with the JAC-2D code (22) running on a Cray YMP computer. The JAC-2D code is a fully vectorized code which is designed to examine nonlinear stress analysis problems, and utilizes the conjugate gradient technique. This code is capable of incorporating a wide variety of constitutive relations for the various materials specified in a given mesh geometry. For all of the FEA results presented in this paper, the alumina ceramic, constituting the body of the area seal, deforms according to an elastic constitutive law, while the

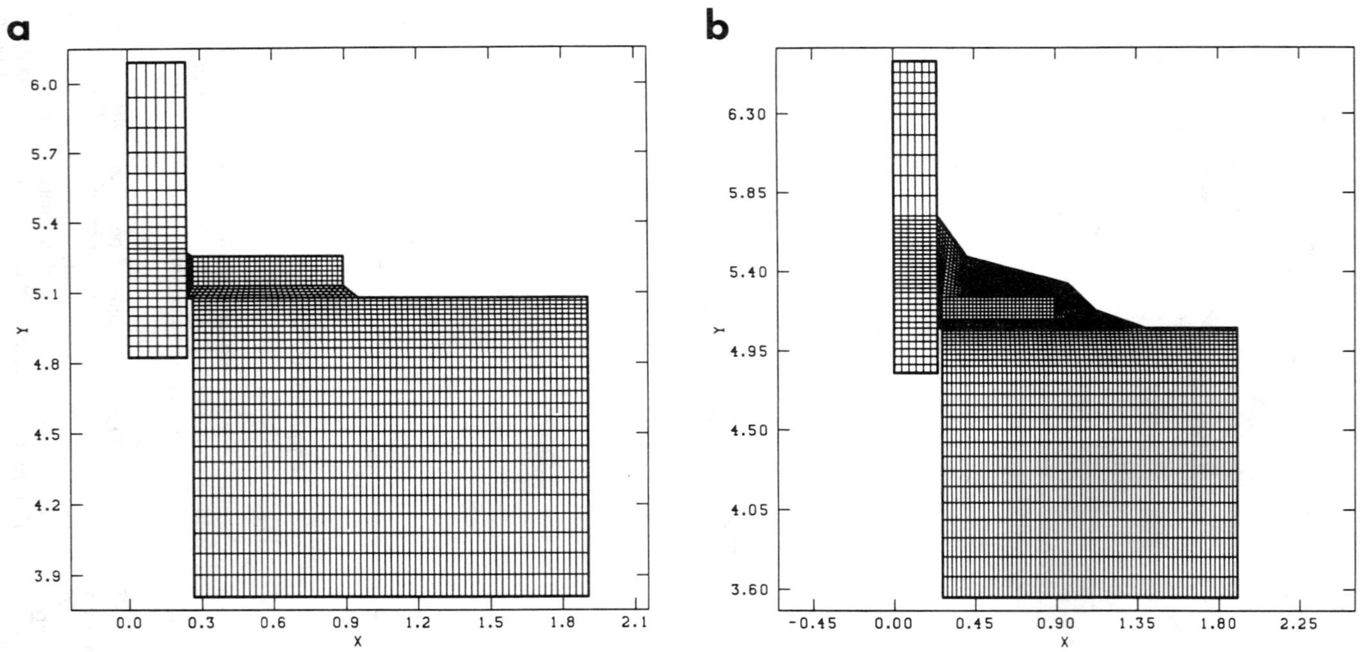

Figure 9 - Sketches of the two different area seal geometries used for FEA calcualtions. Both schematics have symmetry axes along the line x=0; scale in millimeters. (a) Area seal geometry #1: no Cu brazement on top of the Kovar washer. The line x = 0 is the axis of symmetry. (b) Area seal geometry #2: in addition to the Cu brazement found in (a), Cu is also on top of and beyond the end of the Kovar washer.

pin and washer were assumed to be Kovar, which is modeled using an elastic/plastic constitutive law incorporating a temperature dependent yield stress and a linear strain hardening relationship. All of the assemblies were assumed to be in a stress-free condition at the beginning of the cooldown, which was taken to be the melting point of Cu, 1084°C.

In order to illustrate the effect of the type of constitutive law used for the Cu brazement on the residual stresses in the alumina ceramic, five different constitutive laws were assumed for the Cu brazement. Those various constitutive relations are listed here in order of increasing complexity, please note that the first two listed are used for time independent calculations: (1) temperature- dependent elastic model, (2) temperature-dependent elastic/plastic model, (3) "high temperature power law creep" model (eq. (4)) over the entire temperature cooldown range, (4) "two power law" model consisting of the high temperature power law creep equation and the intermediate temperature power law creep equation (eq. (6)), and (5) "high temperature power law plus intermediate temperature sinh law" model consisting of the high temperature power law correlation coupled with the intermediate temperature Garofalo sinh law (eq. (9)). In the case of the latter two calculations, the two creep equations were treated as independent mechanisms. Thus, the transition between the high temperature and intermediate temperature creep laws was obtained by incorporating an algorithm into JAC-2D whereby both the high and intermediate temperature correlations would be calculated for a given stress/temperature combination, and the higher of the two predicted minimum strain rates would be provided to the code. This is necessitated by the fact that the transition between the two regimes is a function of both temperature and stress.

The geometries which were examined for this study were both axisymmetric, with the center of the area seal pin being the axis of symmetry - they are shown in Figure 9. Geometry #1, shown in Figure 9a, represents the minimum amount of braze material which could be used to fabricate the area seal, without any Cu on top of the Kovar washer; while Geometry #2 (see Figure 9b) represents the case where substantially more Cu brazement is used, and in addition to the locations in Geometry #1, Cu is present both on top and beyond the end of the Kovar washer. The 0.2% offset yield stress as a function of temperature for Kovar, Palco and recrystallized Cu are shown in Figure 10. While other pin materials such as Palco have been used for area seal designs, we will only discuss results where the pin is Kovar - this is because for both geomteries the pin is very far from the positions of maximum stress in the alumina at the end of brazement. Variations in pin material do not substantially effect the FEA results obtained for the maximum stress in the alumina.

Although it is beyond the scope of this paper to give a detailed description of the JAC-2D code, the following is a very brief summary of how the stresses are obtained for a given run. For a given time step, the temperature is updated according to the inputted thermal profile, which is assumed to be uniform over the entire area seal. For the new temperature, the code first computes the thermal strains for each element, and uses these thermal strains to update the strain tensor. Next, the stress tensor for each element is updated from the strain tensor. This is followed by computation of the incremental strain tensor at each point given the stress tensor and the applicable constitutive material law which applies for a given element. The incremental strain tensor is then added to the existing strain tensor, and the code is then ready to proceed to a new time/temperature increment. The code continues on in this manner, until the the entire range in temperature from 1084 to 23°C is completed.

Results for Geometry #1: Cu Brazement beneath the Kovar Washer and Between Washer and Pin

The first set of FEA calculations examined the effect of the standard 1700 second cooldown cycle on the stresses obtained in the alumina ceramic for the area seal geometry #1. The maximum principal stress in the alumina ceramic is plotted as a function of temperature during the cooldown cycle in Figure 11. A number of interesting features are apparent in this figure. First, the temperature dependent elastic model for Cu is clearly not able to predict stresses, as the stresses in the ceramic obtained with this model are extremely large within 250°C of the Cu melting point. The other time-independent model, the elastic-plastic model, is clearly more reasonable, and predicts a peak in the maximum stresses at 400°C, followed by slightly lower levels of stress at temperatures less than 400°C. This decrease in stress below 400°C is

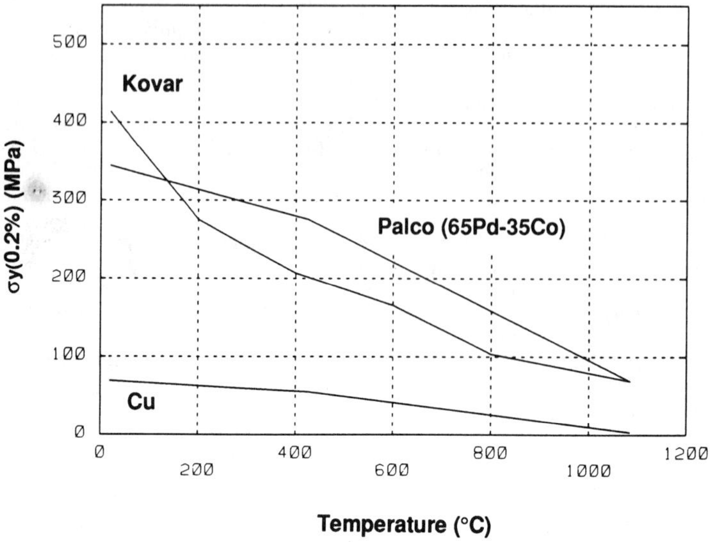

Figure 10 - The 0.2% offset yield strength as a function of temperature for annealed Cu, Palco alloy (65 wt.% Pd - 35Co) and Kovar alloy which was used for the elastic/plastic FEA runs.

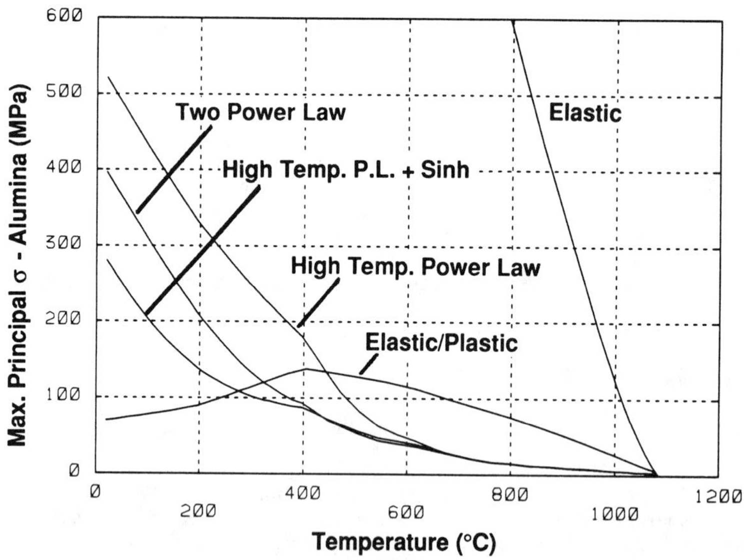

Figure 11 - Results for area seal geometry #1 showing the predicted maximum principal stress in the alumina as a function of temperature using various constitutive relations for the Cu brazement. A total cooldown time of 1700 seconds was used for all the creep calculations.

apparently driven by the fact that the thermal expansion coefficient of Kovar is actually less than that of alumina at these temperatures (see Figure 8a).

The results obtained using the various creep constitutive models shown in Figure 11 predict residual stresses in the alumina ceramic which are substantially less than the elastic-plastic model predictions at temperatures in excess of 600°C. Below 600°C, the predictions are roughly a function of the complexity of the model which was used for a given calculation. The high temperature power law creep model, which incorporated Cu creep data from 1058 down to 589°C, clearely predicts much higher maximum stresses than the other two creep models below 600°C. This is understandable insofar as the higher activation energy in eq. (4) predicts substantially lower creep rates at lower temperatures than either eq. (6) or eq. (9), and therefore underestimates the amount of creep induced stress relaxation which can occur. At intermediate temperatures less than 400°C, the high temperature power law plus sinh law model predicts substantially lower stress levels than the two power law model. Finally, at temperatures less than 275°C, the high temperature power law plus intermediate temperature sinh law model predicts stresses which are higher than the elastic/plastic constitutive model.

A second set of calculations were run with area seal geometry #1, by using the high temperature power law plus intermediate temperature sinh law model and varying the total cooldown time from 1084°C to room temperature. The purpose of these calclulations was to examine explicitly the effect of increased time (and therefore additional creep deformation) in decreasing residual stresses in the ceramic for this particular geometry. To perform these calculations, the cooling curve shown in Figure 3 was simply scaled in time by a multiplicative factor in order to obtain the longer cooldown cycles. The results, shown in Figure 12, indicate that there is indeed an effect on the predicted stresses, but that the decrease in maximum stress is much smaller than the difference between the various creep model predictions shown in Figure 11. This suggests

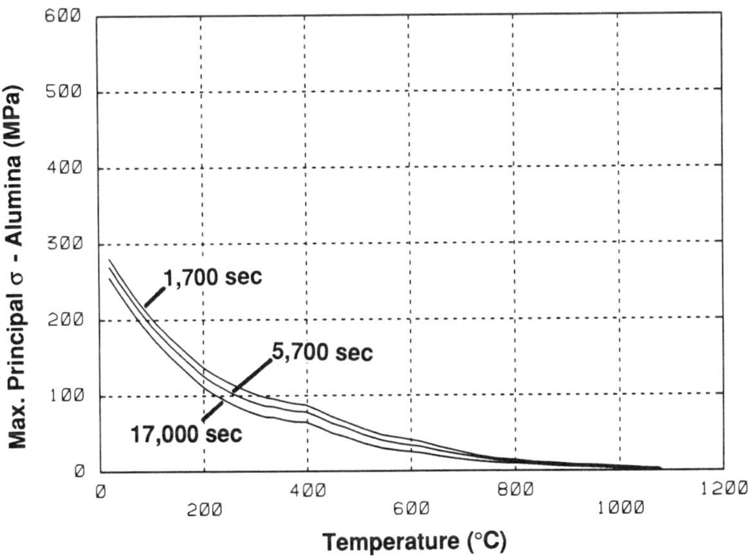

Figure 12 - The effect of total cooldown time from the Cu melting point (1084°C) on the maximum principal stress in the alumina using area seal geometry #1. All calculations shown used the high temperature creep law (eq.(4)) coupled with the intermediate temperature Garofalo sinh law (eq. (9)) as the constitutive relations for the Cu brazement. The cooling profiles for the two longer times simply scale the original cooldown curves in time by factors of 3.35 and 10, respectively.

that the creep kinetics for Cu at these temperatures have become quite sluggish, at least when one uses a model based on minimum creep rate correlations. On the other hand, it is interesting to note that differences in the predicted stresses at elevated temperatures are difficult to resolve above 750°C, suggesting that for the levels of stresses present in the brazement, the creep kinetics for unalloyed Cu during cooldown from 1084 to 750°C are extremely rapid.

Results for Geometry #2: Cu Brazement Also Above and Extending Past the Kovar Washer

A final set of calculations compared the predicted stresses for the elastic plastic model with the high temperature power law plus intermediate temperature sinh law model for the case of an area seal geometry with additional Cu braze material present. The geometry used has been termed "area seal geomertry #2" and is illustrated in Figure 9b. The standard 1700 second total cooldown cycle shown in Figure 3 was used for these calculations, with the predicted maximum principal stress in the alumina plotted as a factor of temperature in Figure 13. The results are similar to the results obtained with these two models for geometry #1, except that at temperatures less than 300°C slightly higher maximum stresses in the alumina ceramic are predicted by the high temperature power law plus intermediate temperature sinh law model. The net effect is to shift the crossover point where the creep model predicts higher residual stresses than the elastic/plastic model to a higher temperature. Note also that the elastic/plastic model predicts lower stresses, and no peak in stress near 400°C as in the case of Geometry #1. This is apparently due to the larger volume of copper in the present geometry, thus the contribution of the Kovar to the residual stresses in the ceramic is proportionately smaller.

Discussion

The difference between the maximum stresses predicted by the high temperature power law plus intermediate temperature sinh law model for the two area seal geometries suggests that it is generally more desireable to minimize the amount of braze material present on the area seal.

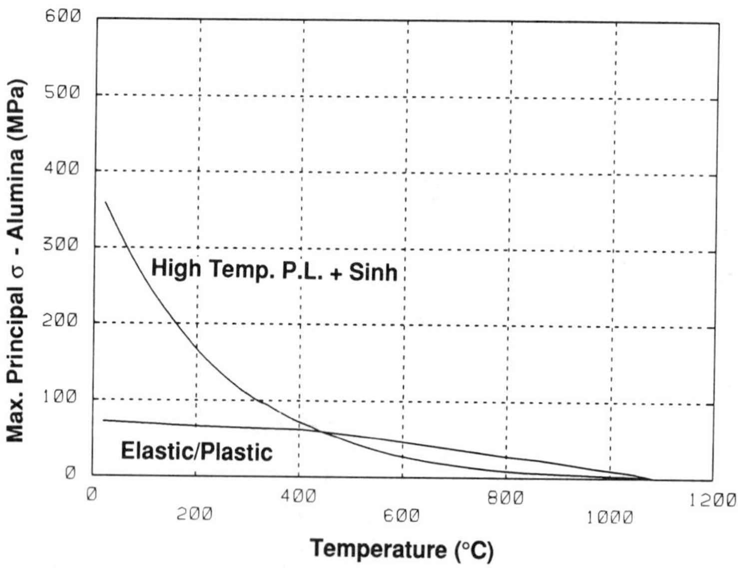

Figure 13 - Finite element analysis results obtained for area seal geometry #2 (see Figure 9b). Note that the crossover between the "High Temperature Power Law + Intermediate Temperature Sinh Law" results with the Elastic/Plastic results occurs at a higher temperature than for area seal geometry #1 (see Figure 11). Total cooldown time of 1700 seconds was used for the creep model.

generally more desireable to minimize the amount of braze material present on the area seal. This consideration is often countered by other mechanical requirements on the seal, such as tensile pull strength, which normally drives one in the direction of increasing the amount of braze material present.

The results of this study raise the important question, "What can be done to improve the creep constitutive models in order to obtain more reasonable stress predictions at intermediate and low temperatures?" One apparent limitation of the current study is that we have not accounted for primary creep behavior at low and intermediate temperatures. Inclusion of primary creep behavior would undoubtedly lead to lower predicted stresses because a substantial portion of the brazement undergoes significant amounts of creep strain. This is illustrated by the contour plot of effective strain (23) obtained for the high temprature power law plus intermediate temperature sinh model results shown in Figure 14. With a significant amount of the brazement undergoing effective creep strains in excess of 4%, an unrealistically high buildup in stress in these areas occurs because we have used a minimum creep rate law. Thus, for these high creep strain regions, a primary creep model would mean that creep would occur at substantially faster strain rates than with the present minimum creep rate correlations.

Raj and Langdon have shown (13) that the ratio between the minimum creep rate and the primary creep rate for Cu becomes quite large (often greater than a factor of 20-30) at temperatures less than 550°C, based on their results for applied stresses of 25 and 62 MPa. At higher temperatures, they found that the ratio between these two creep rates is almost always less than an order of magnitude. Thus, as in the case of FEA calculations for soldered

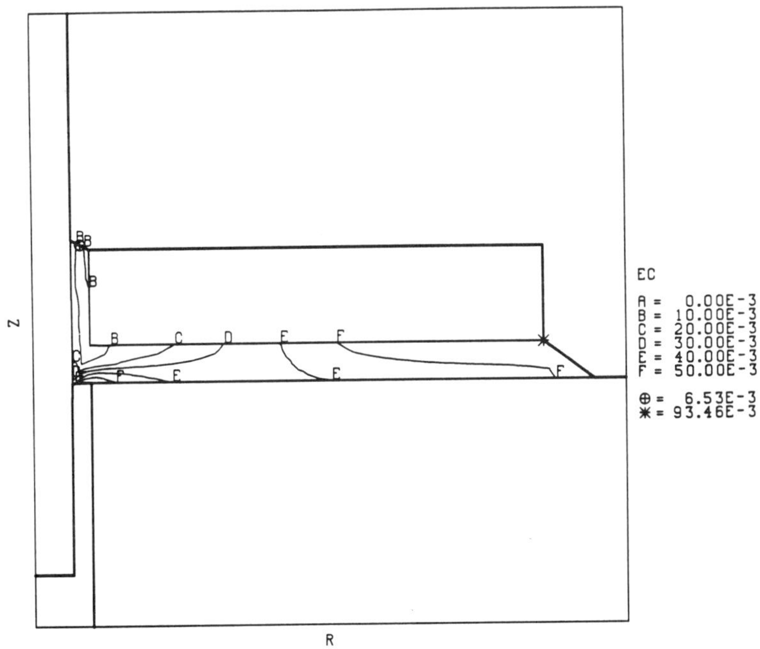

Figure 14 - Contour plot showing the levels of effective creep strain obtained after the normal cooldown cycle (1084 to 23°C in 1700 sec) for area seal geometry #1. The "High Temperature Power Law + Intermediate Temperature Sinh Law" constitutive model was used for the Cu brazement. Note that the maximum effective strain occurs at the corner of the brazement which touches the Kovar washer, while the minimum point is found at the top surface of the brazement between the pin and the washer.

electronic assemblies (1,2) discussed earlier, there is probably little difference in the stresses which would be predicted with more complicated high temperature correlations for Cu which may incorporate primary strain behavior. It appears that future work on creep constitutive relations for Cu at low and intermediate temperatures should include primary creep behavior in order to achieve more realistic residual stress predicitions.

Conclusions

1. Of the three minimum creep rate models used for the unalloyed Cu brazement, the model which incorporates the high temperature power law plus intermediate temperature sinh law fits the creep data over the widest range of temperatures, and appears to yield the most reasonable maximum stress predictions using FEA. This particular model incorporates a high temperature power law, eq. (4), along with an intermediate temperature Garofalo sinh law, eq. (9). At high temperatures, the creep of the Cu brazement is quite rapid, resulting in low residual stresses. However, the maximum stresses in the ceramic become quite high at temperatures less than 250°C.

2. For the case of FEA calculations which incorporate time-independent constitutive relations for the Cu brazement, the elastic/plastic correlation appears to yield more reasonable stress predictions across the entire range of temperatures included in the cooldown cycle.

3. The creep models do not presently include primary creep behavior at intermediate and low temperatures, and incorporation of primary creep behavior would undoubtedly lead to calculations which would predict lower residual stresses in the alumina ceramic.

References

1. H. S. Morgan, "Thermal Stresses in Layered Electrical Assemblies Bonded with Solder," paper #90-WA/EEP-14 presented at the 1990 ASME Winter Meeting (available from ASME, 345 E. 47th St., New York, NY 10017).

2. R. J. Bourcier and J. J. Stephens, "A Numerical and Experimental Study of Residual Stresses and Thermal Fatigue in Soldered Electronic Assemblies," Computer Modelling of Fabrication Processes and Constitutive Behavior of Metals, ed. J. J. M. Too (Ottowa, Canada: Canadian Government Publishing Centre, 1986), 513-526.

3. P. O. Charreyron, D. O. Patten, Jr., and B. J. Miller, "Modeling of Ceramic to Metal Brazed Joints", Ceramic Engineering and Science Proceedings, 10 (11-12) (1989) 1801-1824.

4. J. J. Stephens, S. N. Burchett and F. M. Hosking, "High Temperature Compression Creep of Eutectic Ag-Cu Alloy and Application to Metal/Ceramic Joining," The Metal Science of Joining, ed. M. J. Cieslak (Warrendale, PA: TMS, 1991) (in press).

5. O. T. Iancu, D. Munz, B. Eigenmann, B. Scholtes and E. Macherauch, "Residual Stress State of Brazed Ceramic/Metal Compounds, Determined by Analytical Methods and X-ray residual Stress Measurements," Journal of American Ceramics Society, 73 (5) (May, 1989), 1144-1149.

6. K. Suganuma, T. Okamoto and M. Koizumi, "Influence of Shape and Size on Residual Stress in Ceramic/Metal Joining," Journal of Materials Science, 22 (1987), 3561-3565.

7. J. J. Stephens et. al., Internal Memorandum to J. A. Wilder, Sandia National Laboratories, Albuqueruqe, 8 November, 1990.

8. B. L. Jones and C. M. Sellars, "The Creep of Copper, Copper-10 at.% Nickel and Copper-10 at.% Gold Alloys," Metal Science Journal, 4 (1970), 96-102.

9. K. Monma, H. Suto and M. Oikawa, "High Temperature Creep of Nickel-Copper Alloys," Journal of Japanese Institute of Metals, 28 (1964) 253-258.

10. J. J. Stephens and F. M. Hosking, Internal Memorandum to R. W. Fisher, Sandia National Laboratories, Albuquerque, 2 April, 1990.

11. C. R. Barrett and O. D. Sherby, "Steady-State Creep Characteristics of Polycrystalline Copper in the Temperature Range 400 to 950°C," Transactions of the Metallurgical Society of AIME, 230 (10) (1964), 1322-1327.

12. E. C. Muehleisen, C. R. Barrett and W. D. Nix, "On the High Temperature-Low Stress Creep of Copper," Scripta Metallurgica, 4 (12) (1970), 995-999.

13. S. V. Raj and T. G. Langdon, "Creep of Copper at Intermediate Temperatures - I. Mechancial Characteristics," Acta Metallurgica, 37 (3) (1989), 843-852.

14. C. R. Barrett, J. L. Lytton and O. D. Sherby, "Effect of Grain Size and Annealing Temperature on Steady-State Creep of Copper," Transactions of the Metallurgical Society of AIME, 239 (2) (1967), 170-180.

15. O. D. Sherby and P. M. Burke, "Mechanical Behavior of Crystalline Solids at Elevated Temperatures," Progress in Materials Science, 13 (1967), 325-390.

16. J. Askill, Tracer Diffusion Data for Metals, Alloys and Simple Oxides (New York, NY: IFI Plenum, 1970), 32-33.

17. S. H. Goods, "Mechanisms of Gas Bubble Embrittlement of Metals" (Ph.D. Thesis, Stanford University, 1977), 80-81.

18. F. Garofalo, Fundamentals of Creep and Creep-Rupture in Metals (New York, NY: The MacMillan Company, 1965), 51-55.

19. R. L. Bertolotti, "Thermal Expansions of Kovar and Ceramvar and Seals of These Materials to Alumina" (Report SAND74-8003, Sandia National Laboratories, Livermore, September, 1974).

20. J. H. Gieske, "Temperature Dependence of the Elastic Moduli of Kovar and OFHC Copper" (Report SAND75-0311, Sandia National Laboratories, Albuquerque, December, 1975).

21. "Coors Alumina and Beryllia Properties Handbook" (Coors Ceramic Bulletin #952, Coors Poreclain Company, Golden, Colorado, 1967).

22. J. H. Biffle, "JAC - A Two Dimensional Computer Program for Nonlinear Quasitatic Response of Solids with the Conjugate Gradient Technique" (Report SAND81-0998, Sandia National Laboratories, Albuquerque, April, 1984).

23. O. Hoffman and G. Sachs, Introduction to the Theory of Plasticity For Engineers, (New York: McGraw Hill Company, Inc., 1953), 28.

MATERIALS FACTORS AFFECTING JOINING OF SILICON NITRIDE CERAMICS+

Stathis D. Peteves and Michael G. Nicholas*

Institute for Advanced Materials, Commission of the European Communities, Petten, The Netherlands

Abstract

Reliable techniques are needed for the joining of ceramics, and this paper describes the beneficial roles played by the reactive metals Cr and Ti in the brazing and diffusion bonding of a range of silicon nitride ceramics. Ti and Cr present as alloying elements in brazes or diffusion bonding foils segregate to the ceramic/metal interface and promote joining by the formation of nitride reaction product layers. Results are presented and analysed that describe the effects for Ag, Cu and Ni brazes and for Ni diffusion bonding foils. The influence of reactive specie concentration, surface modification of the ceramic and environment on the microstructure and mechanical integrity of joints is described.

+ Part of the research was carried out within the Commission R&D programme.

* On leave from AEA Technology, Harwell Laboratory, UK.

Metal - Ceramic Joining
Edited by P. Kumar and V. A. Greenhut
The Minerals, Metals & Materials Society, 1991

Introduction

New ceramic materials have been developed in recent years with properties that can be exploited in high temperature, highly stressed advanced engineering projects. Amongst these, silicon nitride has high temperature strength/weight ratios that are superior, for example, to those of nickel base superalloys. However, successful exploitation of such advanced engineering ceramics can depend on the availability of reliable ceramic joining processes, one of the identified key enabling technologies(1). Thus the manufacture of complex ceramic components within tight engineering tolerances can be facilitated by bonding together geometrically simple shapes. Similarly, the attractive properties of ceramics can be exploited economically by using them as inserts bonded to crucial areas of otherwise metallic structures. Processes as varied as fusion welding and adhesive bonding have been used to join ceramics(2), but diffusion bonding and brazing have attracted particular attention for making joints between non-oxide ceramics, between dissimilar ceramics, and between ceramics and metals.

Brazed joints are produced by the flow of molten metals or alloys into narrow gaps between workpieces, but diffusion bonded joints are formed by the application of pressure to bring the workpieces into maximum contact. Very often, a ductile metal foil is interposed between the workpieces to be diffusion bonded so that the basic structures of brazed and diffusion bonded joints are essentially similar. The properties of these joints depend on their size and geometry, the bulk characteristics of the parent and filler materials, and the physico-chemical conditions achieved at the filler/workpiece interfaces.

Forming high quality interfaces between metals and ceramics can be difficult because of their differing physical properties and lattice binding characteristics. Silicon nitride for example is 70% covalent(3) and hence is short of the delocalised electrons that bind metal lattices and could form a continuum across metal/ceramic interfaces. To form high quality interfaces it is usually necessary to change their chemistry by reactions to produce bridging compounds that have lattice binding characteristics compatible with both the metal and the ceramic.

This is often achieved in practice by using chemically reactive metals such as Al as brazes or diffusion bonding foils(4,5), but the relatively low melting point of this metal limits its usefulness for many demanding high temperature applications. Another approach is to promote bonding by using alloy brazes or foils that contain reactive components.

The present paper, will discuss the roles of Ti and Cr additives to Cu, Ag, and Ni brazes, and to Ni diffusion bonding foils utilized to form Si_3N_4 and Sialon joint interfaces. Although attention is focussed on the material factors that affect the quality of interfaces, the effects of specific surface characteristics of the ceramics on the joint properties will be also discussed.

Materials and Techniques

Commercially available Si_3N_4 and Sialon ceramics were used in this work as detailed in Table I. The metals and alloys used were also commercial quality materials as listed in Table II, but small laboratory batches of Cu-Ti and special Ni-Cr alloys were used in some experiments.

Table I Ceramic Materials Employed in this Study

Ceramic	Composition (wt%)	Source
Si_3N_4	$9\%Y_2O_3$ + $1.35\%Al_2O_3$	Ceradyne, USA
"	5% " + 2.9% "	Toshiba, Japan
"	5% " + 3% "	Nippon Steel, Japan
" (Ekasin S)	$6\%La_2O_3$ + $4\%Al_2O_3$	ESK, Germany
Sialon 101	Si_3N_4 + Al_2O_3	Lucas-Cookson, UK
Sialon 201	" + " + Y_2O_3	"

The prime fabrication facilities used to make the metal/silicon nitride joints were two evacuable high temperature furnaces. One had Ta heating elements and radiation shields and was always used under vacuum conditions for the brazing experiments. The brazing temperatures ranged from 1000 to 1150 °C as specified later, but experiments were also conducted at lower temperatures to determine wetting behaviour. At the highest brazing temperature

Table II Metals and Alloys Employed in this Study

Material	Composition (wt%)	Source
Cu	99.9%	Goodfellows, UK
Cu-Ti	2%,5%,7.5%,10%Ti	Laboratory samples
Ag-Cu-Ti	28%Cu-2%Ti	GTE Wesgo, USA
HTN2	Ni-7%Cr-4.5%Si	Wall Colmonoy, UK
Ni-P-Cr	11%P-0%,-14%,-23%Cr	"
Ni-Cr	20%Cr	Goodfellows, UK
Ni-Cr	5%, 10%, 15%Cr	Laboratory samples

of 1150 °C, the vacuum was typically $1x10^{-5}$ mbar with partial pressures of $1x10^{-6}$ mbar for H_2O, CO, CO_2, $6x10^{-7}$ mbar for N_2 and H_2 and $3x10^{-8}$ mbar for O_2. The second processing facility had a W heating element, Mo shields and graphite rams through which uniaxial pressure was applied during diffusion bonding experiments. Typical conditions used were to apply a joining pressure of 100 MPa for 1 h at 1200 °C after the furnace had been backfilled with 1 bar of Ar, whose N_2 content was 50 ppm. For a few experiments, the hot pressing rig was used after evacuation to $1.3x10^{-6}$ mbar with partial pressures of $25x10^{-8}$, $2.5x10^{-8}$, $9x10^{-8}$, and 10^{-9} for CO, CO_2, H_2O and N_2, and O_2 respectively.

The ceramic samples used in the brazing experiments were small slices approximately 15x15x3 mm^3 whose surfaces had been polished to an average surface roughness, R_a, of 0.35 μm. Those used in the diffusion bonding experiments, however, were 15 mm cubes whose mating surfaces were finished to an R_a of 0.4 to 0.03 μm. Before being used in any experiment, the silicon nitride and metal samples were always degreased by ultrasonic agitation in acetone and then dried using hot air blast.

For wetting experiments, ceramic samples were loaded into the furnace and used as substrates for 2 mm diameter, 4 mm high cylinders of Cu or Ag-Cu brazes. The chamber was evacuated to less than $1x10^{-5}$ mbar and heated at about 600 °C/h to just below the braze solidus temperature. The samples were held at this temperature for 30 min to degas and were then heated in steps of 30 °C to the final experimental temperature, with dwells of 10 min at the final and intermediate temperatures. The extent of wetting achieved was measured by direct observation of the

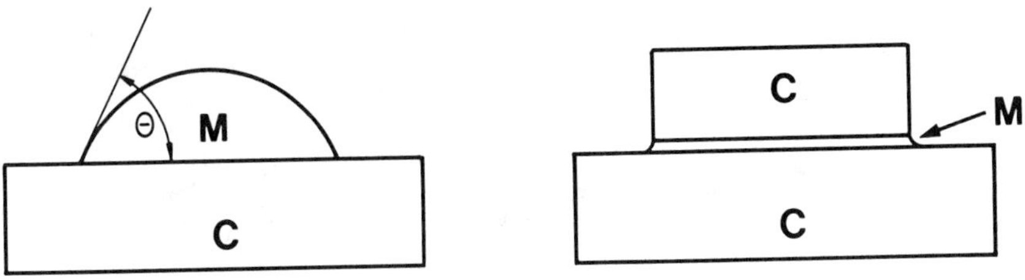

Figure 1 - Schematic of wetting and shear strength samples. (C : ceramic, M : metal).

contact angle, defined as Θ in Fig. 1, using diametrically opposed viewports to provide back illumination and permit photography. Temperatures then already known to cause wetting were used to produce brazed shear strength samples as schematically represented in Fig. 1.

The determination of the wetting behaviour of the Si_3N_4 by the Ni-P brazes containing varying levels of Cr required a modified testing procedure because the braze alloys were available only as powders. In this case, ceramic and braze powders were placed in alumina crucibles and heated up to 1000 or 1200 °C. The samples were subsequently cross-sectioned and macro- and microscopic examination was used to determine the extents of wetting and interaction.

In the diffusion bonding experiments 125 μm thick foils of Ni-20wt%Cr alloy were used as the joining agent. Bonding took place under the conditions described previously and the joints were later machined into 4-point bend test samples (3x4x30 mm^3) as shown in Fig. 2.

For some of these bonding experiments, the mating surfaces of the ceramics had been precoated with reactive species. Ti was applied by vapour deposition and ion-mixing, via dissociation of TiH_2, or by sputter ion plating. On the other hand a 0.1 μm thick layer of Cr was applied by sputter deposition and some of

Figure 2 - As fabricated Si_3N_4/Ni-20%Cr joint and 4-pt bend test samples.

the coated surfaces were subsequently implanted with Xe-ions of 400 KeV energy at doses of $2x10^{16}$ cm^{-2}.
In evaluating the quality of interfaces produced by these various techniques, attention was paid to their strength, microstructure and chemistry. All strength tests were performed at room temperature, with shear strength samples strained at 1 mm/min and bend samples at 0.5 mm/min (20 mm outer span, 10 mm inner span). Microstructures revealed by cross-sectioning and polishing were examined using both optical and scanning microscopy, and interfacial chemistries were determined using electron probe microanalysis(EPMA) and trasmission electron microscopy(TEM).

Results

Brazing

Brazing promoted by Ti. Wetting experiments confirmed that Cu does not wet silicon nitride ceramics at 1150 °C(6), with contact angles of 136° being measured for Ekasin S, and 134° for both Sialon 101 and Sialon 201. However, adding Ti to Cu promotes wetting; 2.5-5% Ti was necessary for Ekasin S, and 7.5-10% for the Sialons at 1150 °C, as shown in Fig. 3. The temperature of

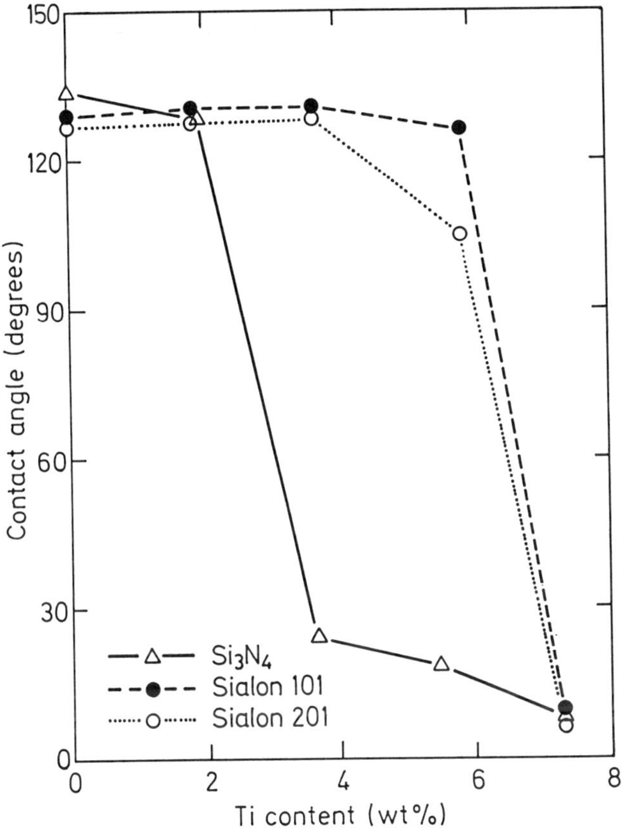

Figure 3 - The effect of Ti additions on the wetting of silicon nitride ceramics by Cu.

1150 °C was chosen as an experimentally convenient upper limit, but it is a significant superheat for the alloys which had liquidus temperatures of 980 to 1040 °C. However, such superheats are not needed, alloys which wetted doing so at or near their liquidus temperatures as demonstrated in Fig. 4, for the Ekasin S ceramic.

Examination of cross-sectioned samples showed that extensive interactions had occurred for both wetted and non-wetted samples as shown in Fig. 5. The thicknesses of the reaction product layers varied, being larger for the interfaces formed by the Ti

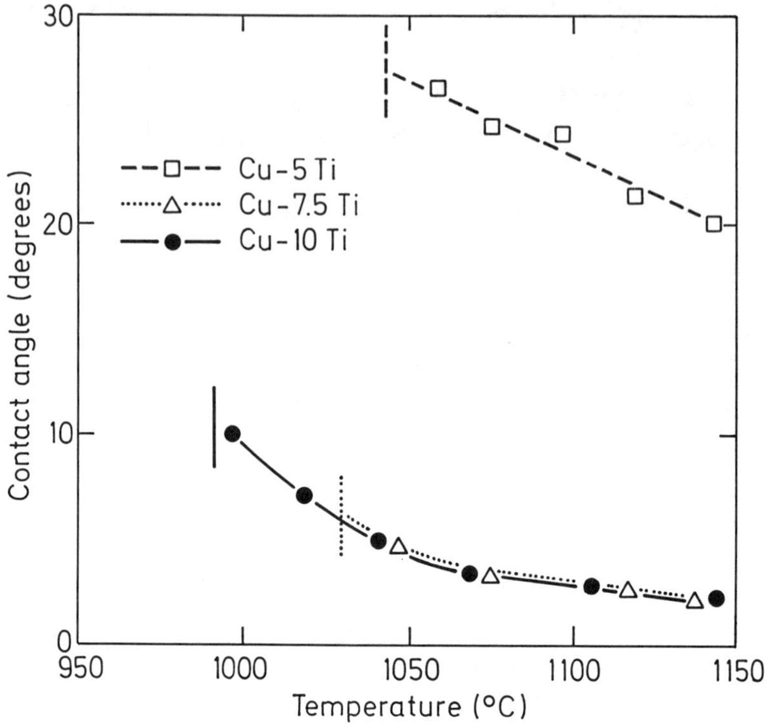

Figure 4 - The effect of temperature on the wetting of Ekasin S silicon nitride by Cu-Ti alloys.

rich alloys and by the Sialons. Further, the thicker reaction product layers were two phase and often cracked.

EPMA surveys, such as those shown in Fig. 6, confirmed that the reaction zone was a two phase product, with that adjacent to the ceramic being Ti rich while that next to the braze alloy was both Ti and Si rich. Although N was not analysed for these samples, other evidence such as hardness and colouration suggested that the reaction products were TiN and Ti_5Si_3.

Binary Cu alloys are seldom used as brazes at present, because of the undesirable stiffening caused by the Ti additions necessary to achieve wetting. Therefore, cosiderable technical attention is focussed on the use of the Ag-28%Cu eutectic alloy, as a solvent, which requires smaller Ti additions than Cu for

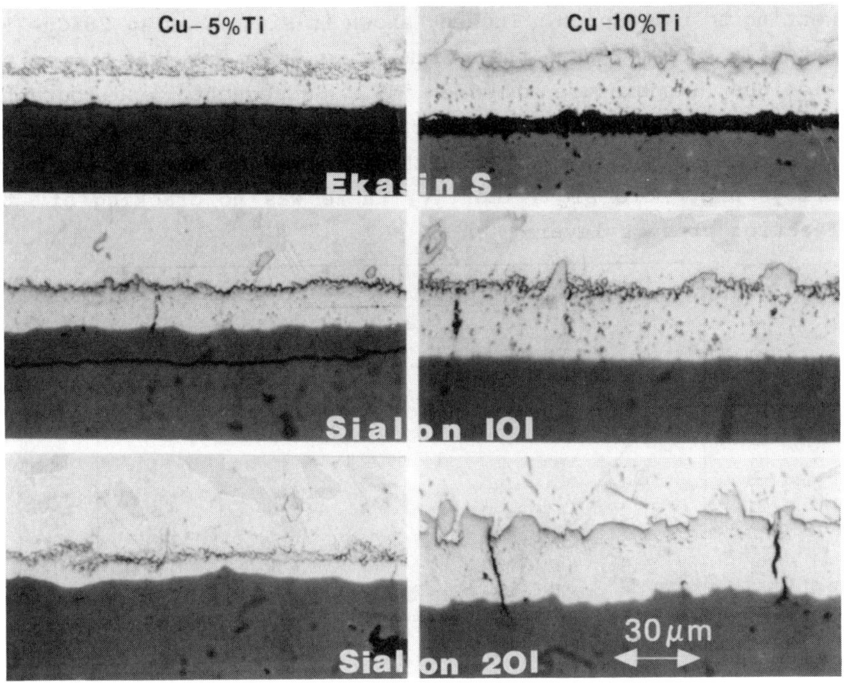

Figure 5 - Interface microstructures of Cu-Ti alloys and Ekasin S, Sialon 101 and Sialon 201 ceramics.

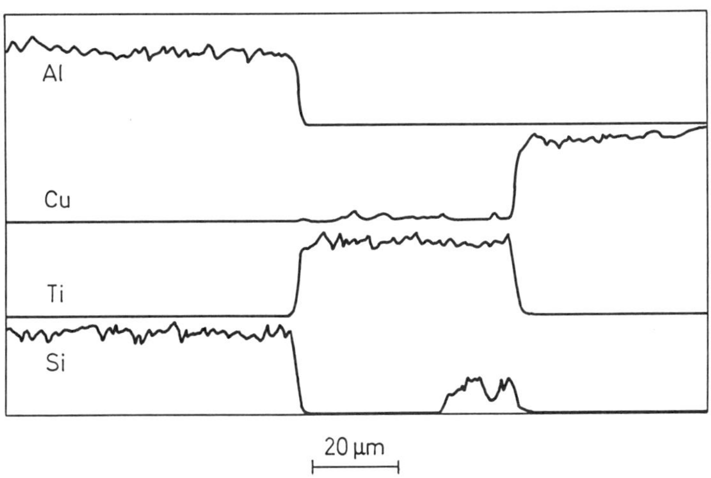

Figure 6 - EPMA elemental traces across the interface formed by Cu-10%Ti with Sialon 201 ceramics.

wetting to be induced. Indeed, shown in Fig. 7, an Ag-28%Cu-2%Ti eutectic alloy wets the ceramics well at or just above the liquidus temperature of 780 °C. Subsequent metallographic examination showed that the reaction product layers were similar in structure but thinner than those formed by the wetting Cu-Ti alloys and, more significantly, there was no cracking of the reaction product layers(7).

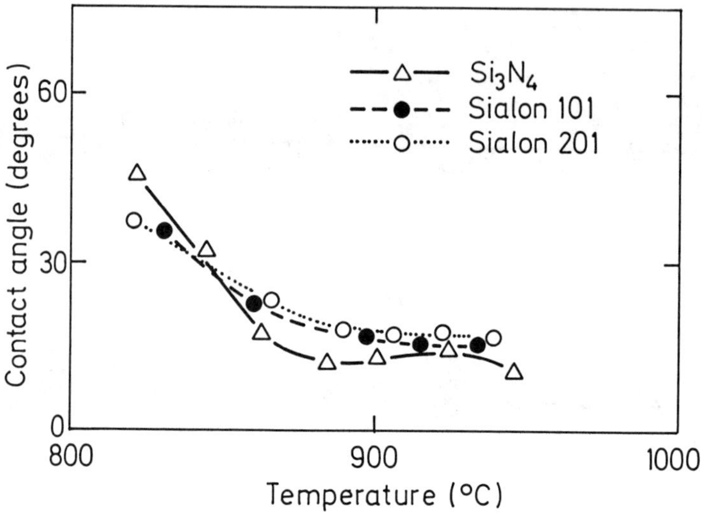

Figure 7 - The effect of temperature on the wetting of silicon nitride ceramics by Ag-27%Cu-2%Ti.

Brazing promoted by Cr. Ni-P-Cr brazes, but not Ni-11%P, wetted the Ceradyne silicon nitride at 1000 °C, and all the Ni-P brazes wetted the ceramic at 1200 °C as shown in Fig. 8. Solidified samples of the Cr-free brazes were porous, and the non wetting interfaces they formed at 1000 °C with the ceramic were not bonded, but all the other solidified brazes were dense and well bonded to silicon nitride. Some erosion of the silicon nitride was caused by contact with the Ni-11%P and Ni-11%P-14%Cr brazes at 1200 °C, but there was no optical evidence of reaction product layer formation.

Bonding promoted by precoating the ceramics. Ti coated Sialon 101 samples were used in wetting and brazing tests with the Ag-28%Cu eutectic and the HTN2 (Ni-7Cr-4.5Si-3Fe-3B) braze alloy as

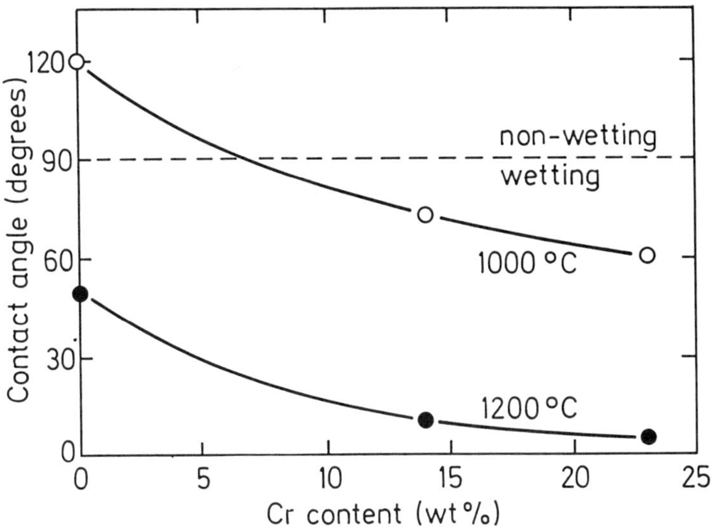

Figure 8 - The wettability of Ceradyne Si_3N_4 by Ni-P-Cr brazes.

filler metals. The contact angle data shown in Fig. 9, show that the Ag-Cu eutectic at 900 °C did not wet the ceramic in the uncoated condition or when thinly coated with 60 nm of Ti by vapour deposition. However, the Sialon ceramic was marginally wetted when coated by the combined deposition/ion-mixing route and well wetted if coated with a 1 µm Ti film by sputter ion plating. The behaviour of the sputter ion plated ceramic was similar to that of uncoated ceramic in contact with the Ag-28%Cu-2%Ti eutectic braze. Work with the Ni braze was less extensive, but tests at 1075 °C showed that it wetted sputter ion plated Sialon very well while not wetting the uncoated ceramic. The sputter ion plated coatings were substantially consumed, but optical microscopy did not reveal any evidence of reaction product layers.

Based on these results, wetting systems were used to produce shear strength samples. Ti sputter ion plated Sialon ceramic brazed with the Ni alloy had strengths of 55-60 MPa. Vapour deposited and ion-mixed Ti coated Sialon brazed with the Ag-Cu eutectic had a shear strength of 45 MPa, somewhat lower than the strength of 58 MPa achieved using the Ag-Cu-Ti braze with the uncoated Sialon. However, the maximum strengths of 75-95 MPa

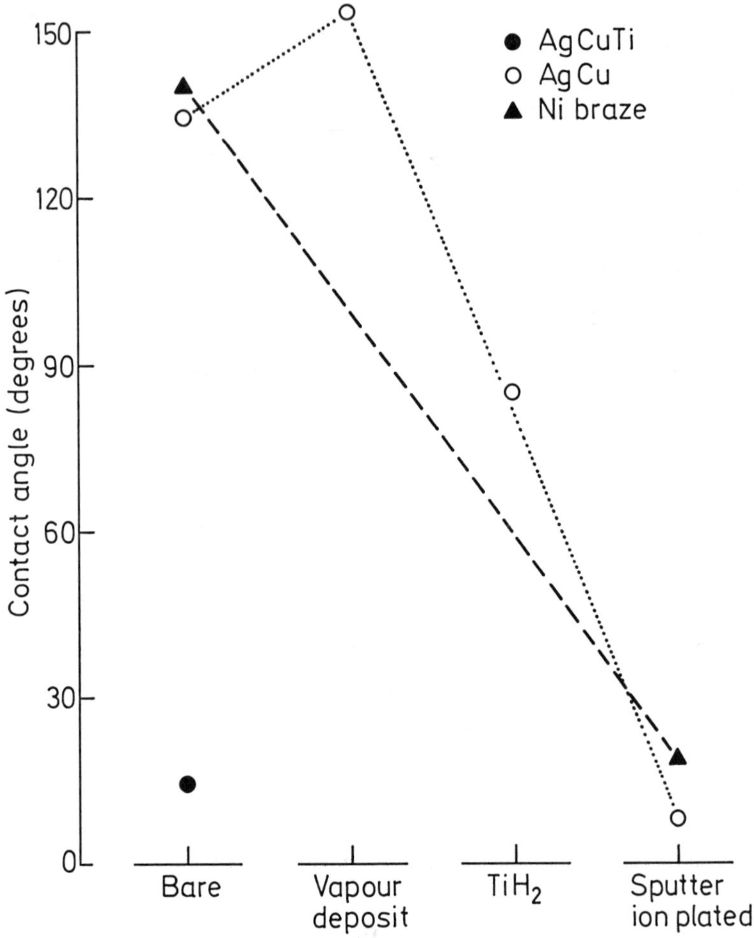

Figure 9 - Wettability data for Ti coated Sialon 101 ceramic.

were measured for the ion sputter plated Sialon ceramic brazed with the Ag-Cu eutectic.

Diffusion Bonding

Bonding of the uncoated ceramics. Earlier experiments using a Ni-20%Cr insert foil (125 μm) to diffusion bond the hot-pressed Si_3N_4 ceramics determined the optimal joining temperature and pressure as, 1200 °C and 50-100 MPa (8). These conditions and an 1 h joining time in an Ar enviroment with a partial pressure of N_2 of $5x10^{-2}$ mbar as the joining atmosphere were adopted as the standard conditions against which the effects of subsequent variations were assessed.

The quality of the joints was assessed by 4-pt bend tests, the results of which were analysed using Weibull statistics; the average flexural strength of the joints, prepared under the standard conditions, was 309 MPa with a 2-parameter Weibull modulus of 2.3, as shown in Fig. 10.

Subsequent optical and scanning electron microscopy revealed that

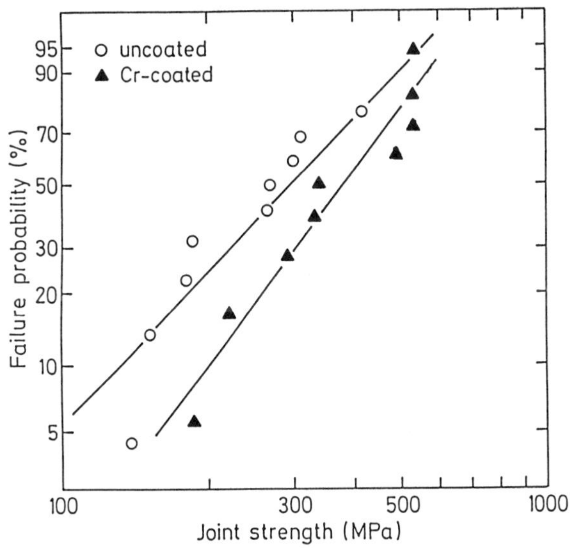

Figure 10 - Weibull plot of Cr-coated and uncoated Si_3N_4/Ni-20%Cr joint strengths.

significant interactions had occurred at the ceramic/metal interfaces during the bonding process. Figure 11 illustrates the growth of a continuous, void free, reaction product layer about 8-10 µm thick at the interface. The interface between this product and the ceramic was quite ragged, but that formed with the foil was relatively smooth and occasionally outlined by micropores as revealed in Fig. 12. Grain boundary precipitation had occurred in the foil at some distance from the interface.

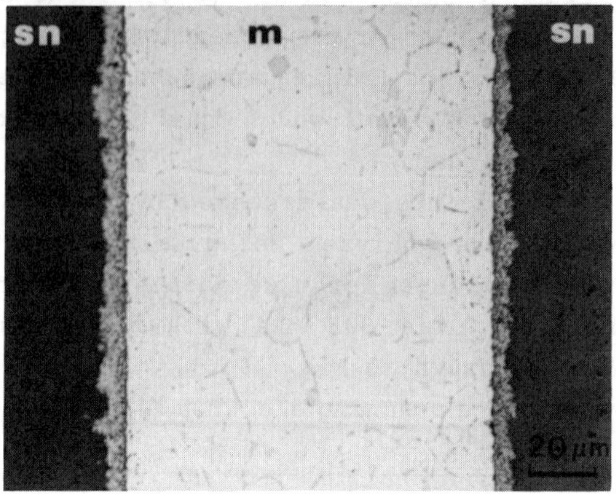

Figure 11 - Optical micrograph of a cross-sectioned uncoated Si_3N_4/Ni-20%Cr joint (sn=Si_3N_4, m=Ni-20%Cr).

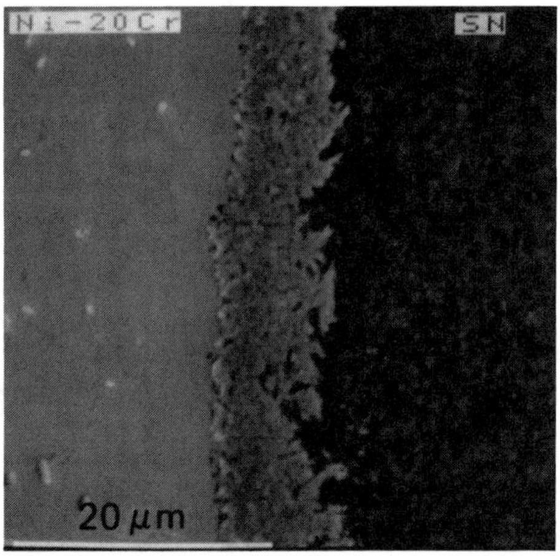

Figure 12 - SE image of a cross-sectioned uncoated Si_3N_4/Ni-20%Cr joint interface.

Wavelength dispersive (WDS) compositional mapping and electron probe microanalysis, such as that shown in Fig. 13, indicated that the reaction product layer and the grain boundary precipitates were Cr and N rich. Additional EPMA quantitative analysis, and X-ray diffraction across the interface showed the composition of the reaction product layer approximated to an

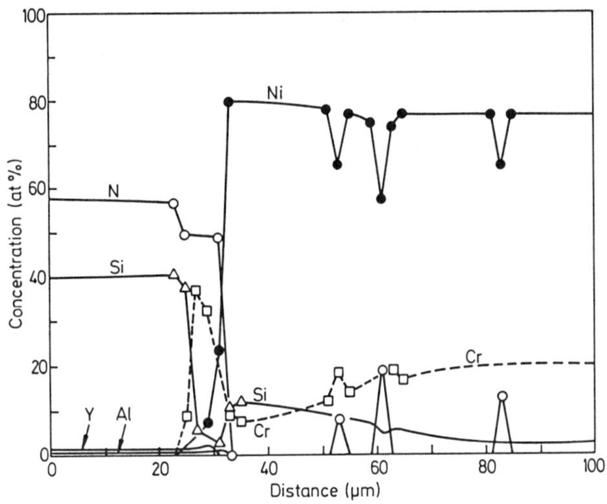

Figure 13 - EPMA scan across an uncoated Si_3N_4/Ni-20%Cr diffusion bonded interface. The scan steps were 1 or 2 μm, and the symbols assigned to each element are for visual clarification only.

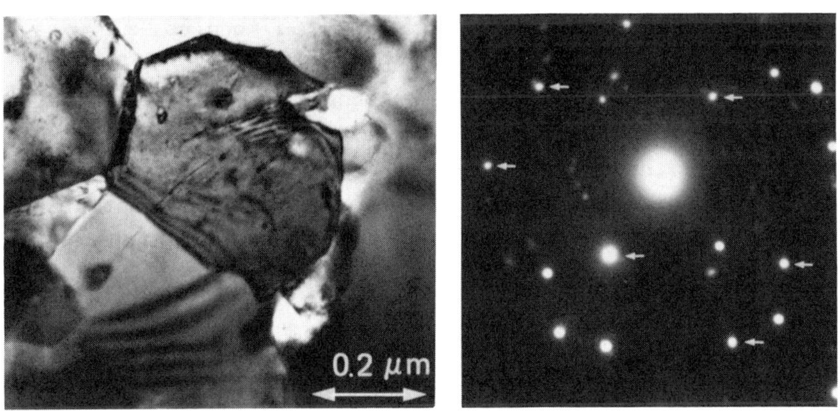

Figure 14 - Bright field TEM image of a CrN grain in the reaction zone and associated electron diffraction pattern. (arrows indicate the CrN spots).

alloyed CrN, while that of the precipitates to Cr_2N, as discussed in more detail elsewhere(9). A typical CrN precipitate and its microdiffraction pattern is shown in Fig. 14. Cr levels were depleted within the foils near the reaction product layers but reverted to those of as received foils over a distance of about 30 μm. Similarly, there was a Si gradient in the foil due to ingress from the ceramic. These Cr and Si gradients were consistent with the operation of a volume diffusion process(9).

Bonding of Cr-coated ceramics. The optimal condition studies revealed that Cr plays an important role in the bonding process. This was further demonstrated by the bonding of 0.1 μm Cr thick sputter coated silicon nitride ceramics; it was realized that the average joint strength and, most importantly, the Weibull modulus were increased compared to those values for the uncoated ceramics as shown in Table III and Fig. 10.

Table III Strength Characteristics* of Si_3N_4/Ni-20%Cr Joints

Silicon Nitride	Uncoated prior to bonding	Cr-coated prior to bonding
Toshiba	σ = 342, m = 1.8	σ = 390, m = 3.4
Ceradyne	σ = 309, m = 2.3	σ = 384, m = 2.93

* average joint strength, σ (MPa), Weibull modulus, m.

Microstructural analyses of these joint interfaces revealed that the Cr coating had a little effect on the interfacial chemistry(9); however, it did cause the Cr and N levels within the reaction product layers to remain relatively constant over their widths, rather than peaking as shown by comparison of Figs. 13 and 15.

Effects of processing variables. Changing the joining environment can have major effects on the interfacial structures. Altering the environment from nitrogenous Ar to vacuum caused a drastic change in the interfacial microstructure, as shown in Fig. 16. The primary reaction product layer was a two layer zone with a thickness about 20 μm. Between this reaction layer and the foil lay a broad two phase zone consisting of irregular islands in a cracked matrix identified as Ni_3Si. The structure of these joints undoubtedly suggest that an inferior joint strength is to be expected.

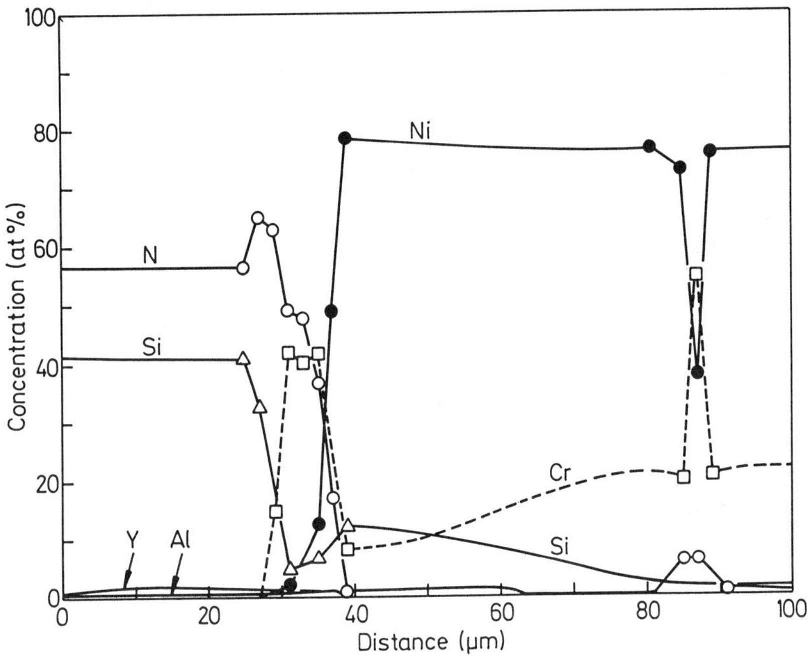

Figure 15 - EPMA scan across a Cr-coated Si_3N_4/Ni-20%Cr joint interface.

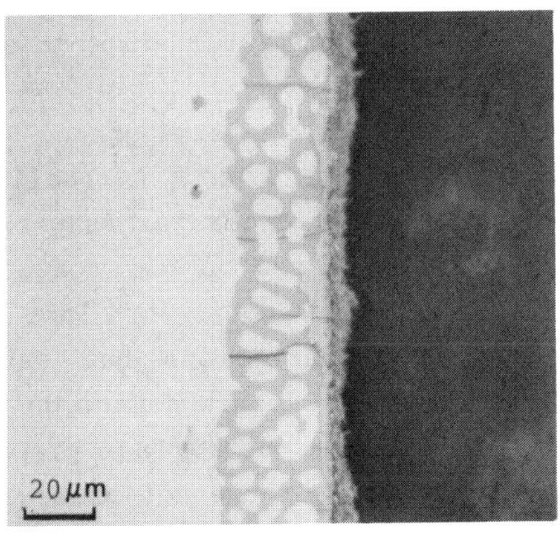

Figure 16 - Optical micrograph of an uncoated Si_3N_4/Ni-20%Cr interface bonded in vacuum.

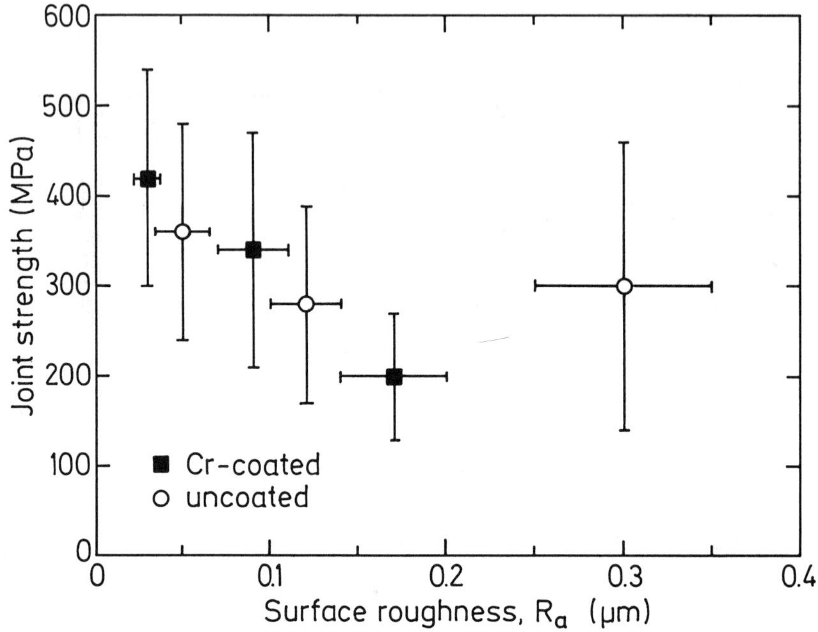

Figure 17 - The effect of the ceramic average surface roughness (R_a) on the Si_3N_4/Ni-20%Cr joint strength.

The average joint strengths for several grades of silicon nitride surface texture are shown in Fig. 17. As realized the smoother ceramic surface results in stronger joints with less scatter in joint strength, for both cases of Cr coated and non-coated ceramics. Clearly these changes were not associated with interface microstructural variations but rather seem to be associated with the residual surface damage on the ceramic prior to bonding caused by the various surface preparation routes. Finally, changing the source of the hot-pressed silicon nitride while still in the Y_2O_3 and Al_2O_3 sintering aids family had an effect on the joint mechanical properties. The average joint strength, but not the modulus, were found to vary proportionally to the parent materials strength as shown in Fig. 18.

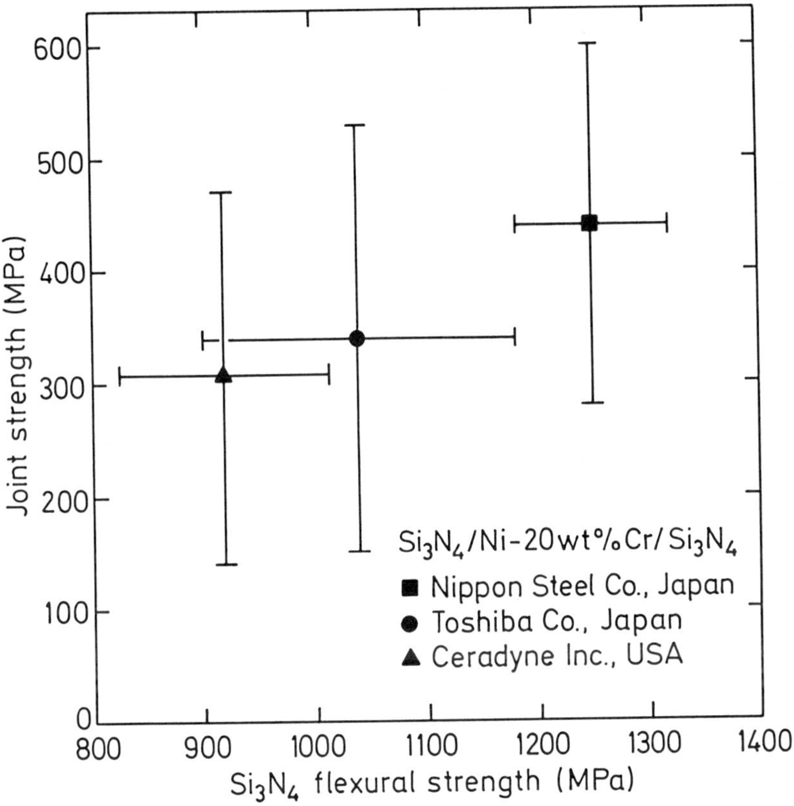

Figure 18 - The average strength of uncoated Si_3N_4/Ni-20%Cr joints bonded in Argon, as a function of the parent Si_3N_4 ceramic flexural strength. (bars indicate ± 1 standard deviation).

Discussion

These studies have demonstrated that both liquid and solid phase bonding of silicon nitride can be promoted by the presence of the reactive species, Cr and Ti. These effects were associated with the formation of metal nitride bridging layers. The initial ceramic/metal interaction, however, will have been a dissolution process with Si and N entering the metal in a ratio of 3 to 4 at a progressively lower rate as saturation levels are approached.

These levels are much lower for N than Si, as shown in Table IV(10), and hence Cr_2N and TiN will be the first reaction products to form despite the lower affinity of Cr for N than Si under standard conditions, as depicted from Fig. 19(11,12).

Table IV Some Si and N Solubility data (at%)

Solute	In Cu at 1150 °C	In Ni at 1200 °C
Si	50	13
N	≈0	≤0.3

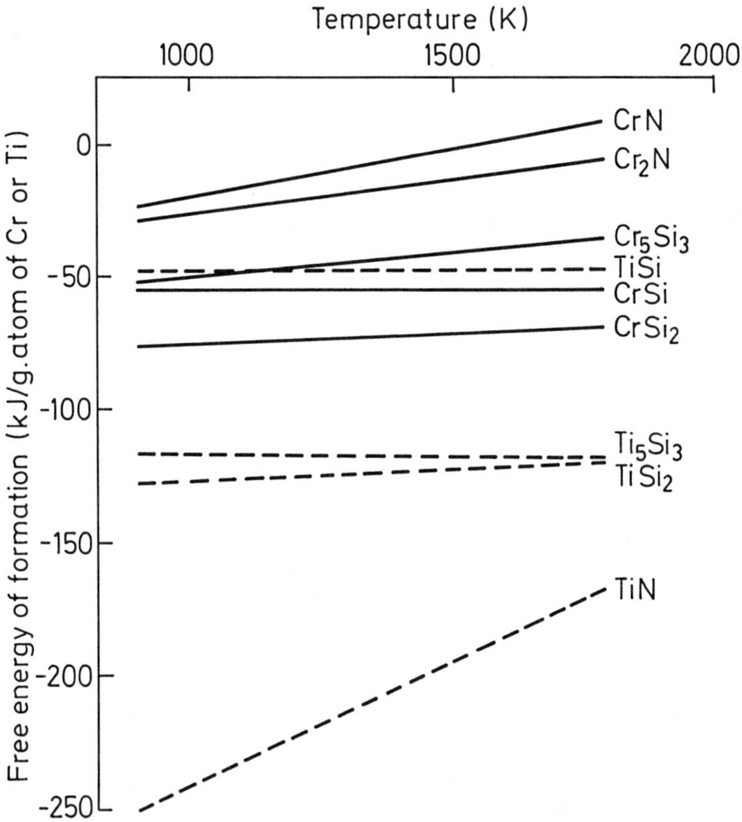

Figure 19 - Free energies of formation for some nitrides and silicides of Cr and Ti.

While generally this sequence will occur, nitrides will not be formed if the Cr or Ti activities are low. In these circumstances, N_2 will be able to escape in the environment or to nucleate bubbles, as observed when Cr lean Ni-Cr foils are used in the diffusion bonding experiments (13). Ultimately silicides will be formed, for example Ti_5Si_3 was produced during brazing experiments to provide an intermetallic coating readily wetted by Cu and Ag-Cu. In the case of diffusion bonded samples, however, the activity coefficient of Si in Ni is so low (14) that silicides form only when the use of a vacuum as the fabrication environment caused the ceramic to disociate and release elemental Si of unit activity at the interface.

While Cr_2N is the most stable nitride of Cr under standard conditions and hence would be the first to form, the reaction product layer present in diffusion bonded samples was an alloyed CrN. This is believed to be the result of a degradation reaction

$$Cr_2N \rightarrow CrN + Cr$$

caused by local depletion of Cr in the foil below the level of about 10% needed to stabilize Cr_2N. This reaction seems to have occurred not only with uncoated samples but also with the Cr-coated ceramics which must have had an initial Cr activity of unity. However the coatings were only 0.1 μm thick and hence would be consumed in the formation of the first 0.15 μm of CrN, a small fraction of the 8-10 μm actually estimated, so once more the principal source of Cr will have been the depleting foil.

By using differing environments, the activities of N available for nitride formation can be varied. In vacuum the partial pressure of N_2, $6x10^{-7}$ mbar in the Ta element furnace and $9x10^{-8}$ mbar in the W element furnace, was so low that the silicon nitride was unstable. However, previous work has shown that the O_2 activity in the Ta furnace, $3x10^{-8}$ mbar was sufficient to remove liberated Si by formation of volatile SiO so that the surface chemistry was not substantially changed (15). In contrast, the O_2 activity in the evacuated W element furnace was only $1x10^{-9}$ mbar so that the ceramic surface should have remained Si coated. The microstructural examinations show that the effect of this was to cause the intial reaction with the diffusion bonding foil to be a siliciding, and enough Si diffused into the foil to form molten Ni_3Si. This penetrated the grain boundaries in the foil to isolate undissolved cores as islands, shown in Fig. 16. While this process assists the achievement of contact between the foil and the ceramics, solidification of the melt on cooling at the end of the bonding process produces a fragile intermetallic and hence detrimental for strong tough bonding.

Finally, it is worth noting the evidence revealed by this study of the connection between microstructures and mechanical properties. With bonding depending on the formation of bridging compounds, it is important that these are not mechanically flawed, cracked, by either mismatched growth strains produced when reactants and product have markedly differing volumes or by mismatched thermal contraction strains generated during cooling. The significance of mismatched strains is well recognised by those concerned with the joining of dissimilar materials, and the use of compliant interlayers has been shown to be beneficial. Thus work reported elsewhere (16) has shown that strong bonds can be formed with Sialon 101 when Al is used as a braze, being both ductile and forming a very thin AlN reaction product layer under optimal joining conditions. In this study, strong brazed bonds were produced when the relatively ductile Ag-27%Cu-2%Ti commercial braze was used, but the strongest bonding was obtained when coated ceramics were brazed with the very ductile Ag-28%Cu alloy. Similarly the excellent plasticity of Ni-20%Cr, which has a room temperature elongation to fracture of about 40% (17), is relevant to the high diffusion bonded joint strengths produced in this work. Even better joint properties may be achievable using more complex interlayers; for example, a ductile metal core first clad with, first, a diffusion barrier to prevent ingress of embrittling species such as Si into the metal, and then a reaction layer that is brought into contact with the ceramic.

In conclusion therefore, this study has shown that the promotion of reactions at ceramic/metal interfaces can enhance wetting and bonding characteristics. Increasing the extent of the reactions progressively improves wetting behaviour. However, the best bonding is associated with a limited degree of reaction, with an excess resulting in the formation of thick fragile interphase product layers. Thus both chemistry and kinetics must be considered when optimising the microstructures for best joint performance during fabrication processes.

References

1. J. Collyear, A programme for the wider application of new improved materials and processes (HMSO, UK, 1985).

2. M.G. Nicholas, in "Designing interfaces for technological applications" ed. S.D. Peteves (Elsevier Applied Science, UK, 1989), 49-76.

3. L. Pauling, The nature of the chemical bond (2nd edition, Cornell Univ. Press, NY, 1948) ,69.

4. K. Suganuma, T. Okamoto and K. Kamachi, J. Mater. Sci., 22(1987), 2702-2706.

5. A. Wicker, P. Darbon and F. Grivon, in Proc. 1st Int. Symp. on "Ceramic components for engines" ed. S. Somiya et al., (Elsevier Applied Science, UK, 1987), 716-720.

6. M. Naka, M. Kubo and I. Okamoto, J. Mater. Sci. Lett., 6(1987), 965-966.

7. M.G. Nicholas et al., "Some observations on the wetting and bonding of nitride ceramics," (Report R 12940, AERE Harwell, UK, 1988).

8. M. Nakamura and S.D. Peteves, J. Amer. Ceram. Soc., 73(1990), 1221-1227.

9. S.D. Peteves, M. Moulaert and M.G. Nicholas, submitted for publication to Metallurgical Transactions (12/1990).

10. M. Hansen, Constitution of binary alloys (McGraw Hill, NY,1958)

11. Smithells Metals Reference Book, ed. E.A. Brandes, (6th edition Butterworths, UK, 1983), 8/24.

12. I. Ansara and B. Sundman, "Computer handling and dissemination of data", ed. P.S. Glasser, (Elsevier, UK, 1987), 154.

13. S.D. Peteves, IAM-CEC, Petten, unpublished work.

14. P. Nash and A. Nash, Bull. Alloy Phase Diagrams, 8(1987), 6-14.

15. M.G. Nicholas and L.M. Jones, J. Mater. Sci. Lett., 8(1989), 265-266.

16. M.G. Nicholas et al., J. Mater. Sci., 25(1990), 2679-2689.

17. Smithells Metals Reference Book, ed. E.A. Brandes, (6th edition, Butterworths, UK, 1983), 22/70.

FRACTURE OF CERAMIC-METAL JOINTS : ZIRCONIA/NODULAR CAST IRON SYSTEM

By

B.S. Majumdar and J. Ahmad
Battelle Memorial Institute, Columbus, Ohio 43201

Key Words: Ceramic, metal, interface, interlayer, bimaterial, fracture, mixed-mode, zirconia, joints

ABSTRACT

Zirconia/nodular cast iron (NCI) joints were successfully fabricated using an active-filler metal approach. Ti in the filler metal primarily was responsible for good bonding. The average strength of the joints was approximately 223 MPa. The toughness of the joints was characterized using the strain energy release rate, G, and its variation with mode mixity, ψ (=$\tan^{-1}K_2/K_1$). For nominally Mode I loading, good agreement was obtained in the value of G at failure (G_c approximately 220 N/m) , when comparisons were made between results obtained from cracked bend bars and cracked diametrally compressed disk specimens. This established geometry independence of a fracture mechanics based approach in predicting failure of cracked ceramic-metal joints. Cracks in the joints were found to propagate partly through the reaction zone of the zirconia and partly through the Ti-rich layer at the braze metal/zirconia interface. Accordingly, the crack was modeled as an interface crack. The strength and toughness of the joints were lower than the base zirconia, and they occurred because of degradation of the zirconia in the reaction zone. G_c was found to be a strongly increasing function of mode mixity, and this was associated with increasing shear deformation of the braze metal immediately adjacent to the zirconia. It is suggested that the strong positive dependence of G_c on ψ be utilized in designing components containing ceramic-metal joints.

1. INTRODUCTION

Recently, there has been significant emphasis on the application of structural ceramics in high-temperature engine components. The approach appears to be to use ceramics at locations experiencing high temperatures and wear, and using conventional metallic alloys at other locations, where the operating conditions are not as severe. This combination of ceramics and metals has resulted in increased importance of ceramic-metal joints, primarily because mechanical

Metal - Ceramic Joining
Edited by P. Kumar and V. A. Greenhut
The Minerals, Metals & Materials Society, 1991

fasteners produce stress concentrations and high contact stresses, which need to be avoided for brittle ceramics. The areas of interest in ceramic-metal joining include the fabrication of strong joints, their mechanical and microstructural characterization and understanding of the mechanisms of strengthening, and development of procedures for designing with ceramic-metal joints. This paper addresses primarily the mechanical and microstructural aspects of a zirconia-nodular cast iron (NCI) joint, which is intended for automotive engine applications.

Similar to monolithic ceramics, the performance of ceramic-metal joints is expected to be controlled by fast fracture initiating from a critical flaw. This is because the fracture toughness of ceramic-metal joints are typically low (lower than the ceramic substrate) [1,2], implying at most small-scale plasticity during failure. However, until recently, ceramic-metal joints were designed and characterized primarily in terms of their bend and shear strengths. Probabilistic aspects were introduced by way of the Weibull parameters. While such a strength based design approach is relatively simple, relying solely on strength and Weibull statistics can lead to significant problems. One reason is that thermal expansion and elastic modulus mismatch between the ceramic and metal will almost always lead to stress singularities at locations where the joint intersects a free surface; in certain cases, the stress may be high enough to lead to small-scale plasticity of the joining interlayer, as was determined in our work on zirconia-zirconia and zirconia-NCI joints [3,4]. The size of such zones, although small, can vary depending upon the extent of material removal by machining and grinding operations. Intermediate thermal cycling can also alter the residual stress distribution and the size of such zones, even in a simple geometry such as the bend bar with a butt joint. Consequently, there exists the potential of overestimating component reliability in design, if differences exist in machining and processing histories between laboratory specimens and the actual component.

One way of dealing with this problem is to treat the highly stressed zone as a flaw, and to design based on fast fracture initiating from such a flaw, or an as-processed or a service induced flaw, whichever is calculated to be of critical size. This type of a damage tolerant approach has a built-in precision, provided ofcourse that the size of defects or highly stressed zones can be estimated reasonably well; it may be noted in this regard that fracture toughness is much less sensitive to flaw distribution than is strength. Also, because highly stressed zones are present in joints from thermal mismatch effects, a damage tolerant approach is probably more important for ceramic joints than even for monolithic ceramics. Accordingly, the primary emphasis in this work was the evaluation of fracture toughness of zirconia/NCI joints. An added feature was mixed-mode fracture. This aspect is important in metal-ceramic joints because of modulus mismatch; a loading pattern that is pure Mode I for a crack in a homogeneous material will lead to Mode I-Mode II loading for a crack at a bimaterial interface [5-7]. Thus, whereas Mode I fracture toughness, K_{Ic}, or the critical energy release rate, G_{Ic}, is generally sufficient to define conditions for fast fracture from a flaw in a homogeneous material, in ceramic-metal joints it also is imperative to determine how the energy release rate (G_c) is affected by the mode mixity (ψ), defined as tan-1(K_2/K_1), where K_1 and K_2 are the local Mode I and Mode II stress-intensity factors, respectively, at the onset of fast fracture.

2. EXPERIMENTS

2.1 Materials Joined (Substrates)

The zirconia (ZrO_2) used for the investigation was a magnesia partially stabilized zirconia (Mg-PSZ), Zycron-L, purchased from Zircoa Products, Inc.. The material was obtained in the form of 50.8 mm diameter by 19 mm thick billets. Experiments on the base zirconia provided a bend strength of approximately 589 MPa, an elastic modulus of 202 GPa, and a fracture toughness (K_{Ic}) of approximately 9.2 MPa/m.

The nodular cast iron was obtained in the form of a large block from Sandia National Laboratories, and corresponded approximately to ASTM A536,Grade 60-40-18. The material had an ultimate tensile strength of 437 MPa, a yield strength of 222 MPa, a Young's modulus of 164 GPa, and a fracture toughness of approximately 85 MPa$\sqrt{m}$.

Table 1 provides constitutive properties of the base zirconia, NCI, and the Incusil ABA braze metal.

2.2 Joining Technique

Final joints for property characterization were obtained by brazing the 50.8 mm diameter zirconia billets to 50.8 mm diameter x19 mm thick NCI blocks. The surfaces of the billets were made flat and uniform by grinding to a 220 grit finish using a diamond wheel. The zirconia was further polished using successive grades of diamond paste, up to 9 μm. All billets were thoroughly cleaned with acetone followed by methanol rinse. The braze metal was lightly abraded and thoroughly cleaned before placing between the zirconia and NCI.

Two brazing processes were used for the fabrication of zirconia/NCI joints; the active substrate process and the active filler-metal process. Both techniques rely on a Ti-rich layer at the ceramic

Table 1. Room Temperature Data for Constituent Materials

Constituent Material	**Young's Modulus, E** GPa	**Fracture Strength,** MPa	**Poisson's Ratio,** ν	**GIc,** N/m	**KIc,** MPa√m	α x 10-6 /C
As-Received Zirconia	202	589	0.22	896.4	9.2	10.3
Nodular Cast Iron	163.4	437	0.28	40750.1	85	14.9
Incusil-ABA Braze Metal	79.3	402	0.36	9878.4	30	18.2

interface to obtain a bond. The active substrate process was initially attempted, where the zirconia was coated with the active metal Ti, before being brazed to the NCI using the filler metal, 60Ag/30Cu/10Sn. Although this technique was partially successful with small joints, it was unsuccessful when scaled up to the size of the billets. Therefore, this procedure was discontinued.

All joints fabricated using the active filler-metal approach were successful; the results discussed in this paper were obtained from joints fabricated using this technique. The braze metal was Incusil ABA, with the composition 59.2Ag-27.5Cu-12.1In-1.2Ti (all in weight percent), obtained from GTE Wesgo. The braze metal foil was sandwiched between the zirconia and NCI, and placed in a vacuum furnace operating between $2x10^{-5}$ and 10^{-4} torr vacuum. The sandwich was heated at a slow rate to the brazing temperature of approximately 750 C, held at temperature for approximately 10 minutes, then slowly cooled to room temperature in vacuum. No pressure was applied during joining; the only contact load was that from the weight of one billet over another. Ultrasonic examination showed that the joined billets were free of macroscopic defects. After joining, the thickness of the braze interlayer was found to range between 30 μm and 75 μm. As will be shown later, the active filler-metal, namely Ti, was responsible for providing strong and reliable zirconia/NCI joints. Optical examination of the joint showed that the zirconia (normally yellow) became dark over a length of approximately 0.4 mm from the zirconia/braze metal interface. This zone corresponded to the reaction zone (ZrO_{2-x}) of the zirconia, the zone being depleted of oxygen by Ti at the zirconia/braze metal interface, during the brazing operation.

2.3 Mechanical Characterization

Materials characterized in this investigation included zirconia/NCI joints, the base zirconia, NCI, and the active-filler braze metal. An overall goal of the entire effort, described in detail in Reference [3], was to predict the failure load of cracked specimens based on the constitutive properties of the substrates and the braze metal. It was shown in References [3,4] that this objective was largely met. However, in this paper, we shall focus attention only on the measured properties of the joints and indicate how they were influenced by the type of loading; i.e., we shall not discuss prediction of failure loads, based on constituent properties.

Joint specimens were machined from the 50.8mm diameter joined billets. The geometries of the cracked specimens are shown in Figure 1; in the figure, the width (t_i) of the braze metal (interlayer), compared with other specimen dimensions, appears much larger than actual. The left side of Figure 1 corresponds to the cracked 4-point bend bar and the right side corresponds to the

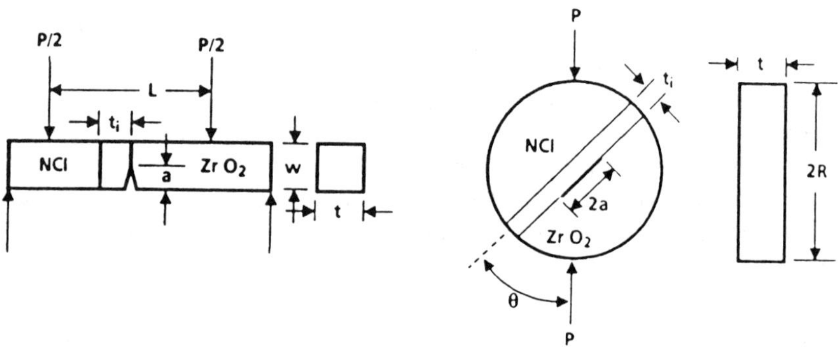

Figure 1. Specimen geometries used for determining fracture toughness of joints and base materials. The left ar right figures correspond to the straight-notched cracked bend-bar and the chevron-notched diametrally compressed disk specimen, respectively.

cracked disk specimen. These two significantly different geometries were selected in order to check the geometry independence of the fracture mechanics based model used in the investigation; the crack-tip region in the bend bar experiences primarily bending loads, whereas the crack-tip in the disk specimen experiences a combination of tensile and shear loads. All testing was conducted at room temperature.

The bend bar, shown in Figure 1, was of depth, W = 5.08 mm, and thickness (or width), t = 5.08 mm. The outer span was 31.8 mm, the inner span was 19.1 mm, and the crack length (a) to W ratio, a/W, was approximately 0.3. The straight notch (approximately 0.2 mm wide) was machined such that the notch-tip was along the braze-metal/zirconia interface. The reason for this location was that uncracked bend bars were found to fail partly along the braze-metal/zirconia interface and partly through the reaction zone of the zirconia. Fatigue precracking was not attempted because earlier experiments, involving toughness evaluation of the as-received zirconia using straight notched and chevron-notched bend bars as well as notched disk specimens, indicated that the straight-notched bend bar provided fracture toughness data that were in excellent agreement with those generated using other specimen geometries. In addition to cracked bend bars, 4-point bend strength testing was performed using uncracked bars with W=2.54 mm and t=50.8 mm, and with the butt joint positioned at mid-span.

The right side of Figure 1 corresponds to the cracked disk specimen. The radius, R, and thickness, t, of the disk were 17.78 mm and 3.6 mm, respectively. The notch (approximately 0.2 mm wide) was machined starting from the two faces of the specimen using a 25.4 mm diameter slitting wheel. This resulted in a built-in chevron notch. The notch was positioned along the braze-metal/zirconia interface, similar to the cracked bend bar. The specimen was oriented at various angles, θ, with respect to the compression loading axis, in order to obtain various combinations of Mode I and Mode II stress intensity factors locally at the crack tip; for a homogeneous material, $\theta=0$ corresponds to pure Mode I loading. For the joined disk specimens with $\theta=0$, the crack grew stably through the chevron-notch region before fast fracture. With increased θ, the crack grew only a fraction of the chevron notch before deviating out of the plane of the notch. That is why, the crack length (2a) at fast fracture was measured after the test; for $\theta=0$, a/R was 0.5, whereas at other angles, a/R ranged between 0.42 and 0.46. Experiments involved loading the disk specimens in compression to failure, and also monitoring the crack mouth opening displacements using a strain-gage type of extensometer.

In addition to the above tests, the stress-strain properties of the braze metal and NCI were determined using tension tests. Also, the thermal expansion of the constituent materials were determined for estimating the residual stresses in the joint.

2.4 Microstructure and Fractography

The microstructure of joints were evaluated using optical and scanning electron microscopy (SEM) techniques. Chemical analysis, using wavelength dispersive spectroscopy (WDS) on a JEOL 733 microprobe, was performed to determine the chemical species responsible for good wetting between interlayer and adherents, and also for evaluating any undesirable reaction product that may have formed as a result of the joining process.

The fracture surfaces of representative specimens were evaluated optically and by using SEM. Limited energy dispersive X-ray analysis (EDS) was conducted to obtain estimates of the chemistry in different regions of the fracture surface. Such observations and measurements helped in identifying the preferred fracture path, namely, whether crack propagated through the adherents, or through the bonding interlayer, or through the interface of interlayer and adherent. Additional confirmation of the fracture path was obtained by sectioning fractured samples perpendicular to the fracture surface, metallographically polishing those sections, and observing such polished sections using optical and scanning electron microscopy techniques.

3. ANALYSIS

3.1 Stresses in Uncracked Bend Bars

The apparent bend strengths of joints were determined from the loads at failure, using the standard bend-bar formula for a homogeneous material. Although such a procedure neglects residual stresses from thermal expansion mismatch, and shielding stresses from modulus mismatch, it provides a basis for comparing the strength of the joint with respect to the strength of the base ceramic, namely zirconia. Weibull analysis was performed on the apparent bend strengths.

Finite element method (FEM) analysis was performed to determine the stress locally at the failure location of uncracked bend specimens; the properties of the constituent materials (substrates and braze metal) were used in the FEM analysis. The following relationships were obtained:

$$\sigma = \sigma'.\sigma_H \quad (1)$$

$$\sigma' = \{1.0+(t_i/W)[1-(2\mu_i/(\mu_m+\mu_c))][(177t_i/W)-10.9]\}\{0.98+\beta_1(2x/t_i)+\gamma(2x/t_i)^2\} \quad (2)$$

for $0.19 \leq \mu_i/(\mu_m+\mu_c) \leq 0.23$.

Here, σ is the actual bend stress on the tensile surface of the bend bar at the bond location, σ_H is the bend stress calculated based on a homogeneous material, and σ' is a shielding factor which accounts for elastic mismatch effects[1]. Also, t_i is the braze thickness, W is specimen depth, μ_i, μ_m, and μ_c are the shear modulus of braze-metal, NCI, and zirconia, respectively, β_1 (=0.039) and γ (=0.0949) are constants, and x is the distance from the center of the braze to the failure location. Since failures occurred essentially at the braze-metal/zirconia interface, x is $t_i/2$. For the interface crack, we obtained σ'=1.04; if failure had occurred at the center of the interlayer (x=0), then σ' would have been 0.91, implying a shielding effect.

The above equations (1) and (2) do not account for residual stress effects. Figure 2 shows the residual stress profile in the zirconia/NCI bend bar, based on FEM analysis of the joint region. Similar stress contours were obtained for other normal and effective stresses, and are documented in Reference [3]. The superposition of these residual stresses, and stresses calculated based on equations (1) and (2), provide actual stresses at failure location for uncracked joined bend bars.

3.2 Cracked Specimens

The fracture mechanics problem analyzed for the zirconia/NCI system is shown schematically in Figure 3. The crack was assumed to be located at the braze metal/zirconia interface, in line with observations of the fracture path of uncracked bend bars.

In the case of interface cracks between two materials 1 and 2, the stress solutions, in polar coordinates are given by [5-7]:

$$\sigma_{22}+i\sigma_{12} = (K_1+iK_2)r^{i\varepsilon}/\sqrt{2\pi r} \quad (3)$$

. . .[1] The terminology "shielding" has been used here because very often the interlayer has lower modulus than the substrates, and this acts to reduce the local stresses inside the interlayer (where failures may occur) compared with a homogeneous specimen.

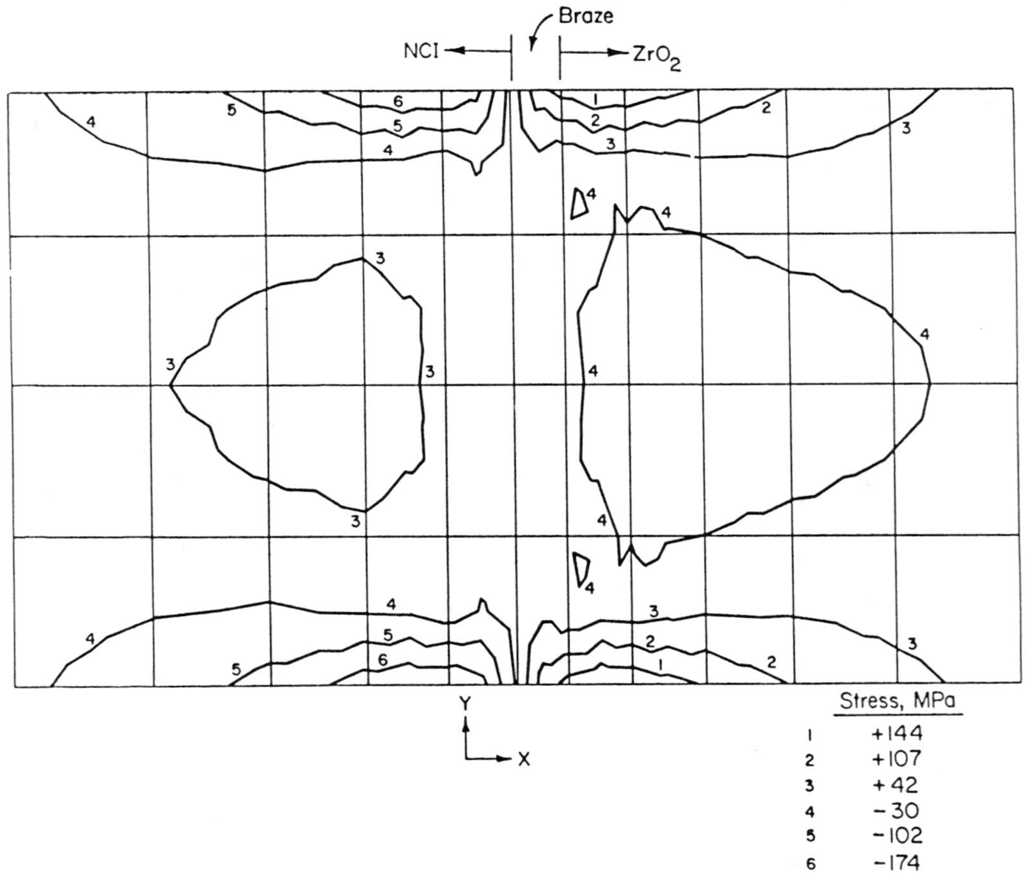

Figure 2. Residual stress (σ_{xx}) profiles for a zirconia/NCI bend bar (uncracked specimen). The y-axis corresponds to [illegible]

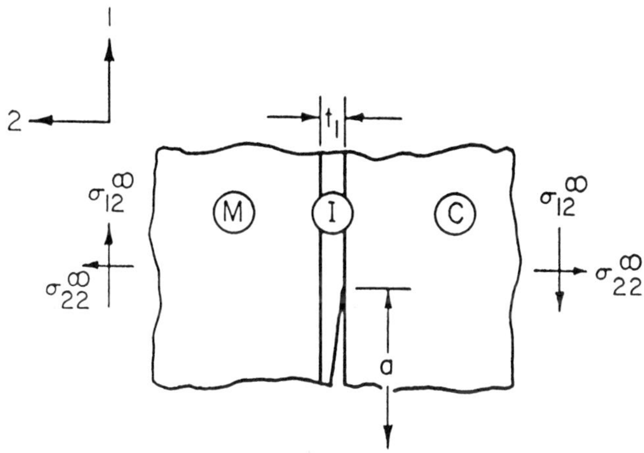

Figure 3. The fracture mechanics problem analyzed in this investigation. The crack location was based on experimental observation of fracture path of an uncracked specimen.

where K_1 and K_2 are the local stress intensity factors, defined to be consistent with homogeneous solutions, and linearly proportional to applied loads. r is the distance from the crack tip, and

$$\varepsilon = 1/2\pi \ln\{(1-\beta)/(1+\beta)\} \quad (4)$$

The phase angle, ψ, between the stress intensity factors is given by:

$$\psi = \tan^{-1}(K_2/K_1) \quad (5)$$

The important Dundurs parameters are:

$$\alpha = (E'_1 - E'_2)/(E'_1+E'_2) \quad (6)$$

and,

$$2\beta = \{\mu_1(1-2v_2) - \mu_2(1-2v_1)\}/\{\mu_1(1-v_2)+\mu_2(1-v_1)\} \quad (7)$$

Here E' corresponds to the reduced modulus (= $E/(1-v^2)$), and v corresponds to the Poisson's ratio. As is evident in equation (3), a length parameter $r^{i\varepsilon}$ enters into the numerator for stresses, making a failure criterion based on the stress intensity factor rather difficult. On the other hand, the energy release rate, G, retains its significance as the amount of energy released per unit of crack growth, and is given by the scaler expression:

$$G = \{K^2_1 + K^2_2\}/(E'_a \cosh^2\pi\varepsilon) \quad (8)$$

and, E'_a corresponds to the average modulus, defined as

$$2/E'_a = \{1/E'_1 + 1/E'_2\} \quad (9)$$

Returning to Figure 3, it has been shown [8] that if the substrate material on either side of the interlayer (i) is the same, and the thickness of the interlayer is much smaller than the crack length, then G is given by the relatively simple expression:

$$G = \{K^2_I + K^2_{II})/E'_S \quad (10)$$

where, now K_I and K_{II} are the stress intensity factors determined for a homogeneous substrate

(S), i.e., based on the same geometry and loading conditions, but assuming that the interlayer did not exist at all. In expression (10), E'_S corresponds to the reduced modulus of the substrate. Additionally, the phase angle is given by:

$$\psi = \tan^{-1}(K_2/K_1) = \tan^{-1}(K_{II}/K_I) + \omega \quad (11)$$

where ω is a relatively weak function of α, and is plotted in Reference [8]. Note that K_1 and K_2 correspond to the local stress intensity factors for the interface crack, whereas K_I and K_{II} correspond to stress intensity factors assuming homogeneous material.

The simplicity of equations (10) and (11) make them attractive for analyzing the fracture behavior zirconia/NCI joints, particularly since the reduced modulus of zirconia and NCI were not significantly different, being 212.3 GPa and 177.3 GPa respectively. In this paper, equations (10) and (11) will be used for calculating the strain energy release rate, and we shall provide results of FEM analysis which support the use of these equations.

Because the substrates (zirconia and NCI) had slightly different moduli, E'_S in equation (10) was obtained by taking their mean, i.e.

$$2/E'_S = 1/E'_c + 1/E'_m \quad (12)$$

where the subscripts c and m refer to the ceramic and metal, respectively. For the zirconia/NCI system, E'_S was calculated as 193.2 GPa, α was 0.36, and ω was obtained as approximately -5 degrees.

We shall rewrite equation (10) as

$$G_A = G_{HS} \quad (13)$$

where G_{HS} is calculated assuming that the entire specimen is composed of homogeneous substrate material. The subscript A has been used in equation (13) to signify that it was obtained by an asymptotic analysis procedure.

FEM analysis was performed on the joined specimens, and included both the cracked bend bar and the cracked disk specimen; residual stress effects were also taken into account. G was calculated using the J-integral approach and it was verified that G was indeed independent of the path. The results were expressed in the form:

$$f^2_s = G_{FEM}/G_{Hi} \quad (14)$$

where now G_{Hi} was calculated assuming that the entire specimen was composed of homogeneous interlayer material. The subscript FEM has been used in equation (14) to signify that it was obtained by FEM analysis. f^2_s corresponded to a shielding factor, and accounts for material inhomogeneity.

Comparing equations (13) and (14), it follows that

$$G_{FEM}/G_A = f^2_s \,.\, G_{Hi}/G_{HS} = f^2_s .\, E'_S/E'_i \qquad (15)$$

the extreme right hand side arising because the stress intensity factors for a homogeneous specimen are independent of the material. Thus, the ratio in equation (15) provides a measure of the error involved in calculating G by using equation (10) and assuming an average E'_S.

For nominally Mode I loading, the following equation was obtained in References [3,4] based on the FEM analysis:

$$f_s = 0.4766 + 1.0468\{\mu_i/(\mu_m+\mu_c)\} \qquad (16)$$

Substituting this, and the elastic properties of the zirconia, NCI, and the braze metal, we obtain:

$$G_{FEM}/G_A = 0.998 \qquad (17)$$

implying excellent agreement indeed. In other words, for the crack lengths involved, G calculated using the asymptotic approach (equation 10) provides a good estimate of the strain energy release rate.

We would like to comment here that the asymptotic solution may provide significant errors if the crack is not a true interface crack. For example, in the case of a zirconia/zirconia joint, fabricated using a ceramic interlayer [3], it was found that failure occurred through the center of the interlayer (whose modulus was approximately 0.55 times that of the substrate). FEM analysis of this crack path, and comparison with equation (10) showed that G_{FEM}/G_A ranged between 0.83 and 0.94, depending upon whether the specimen was the bend bar or the cracked disk, respectively. Thus, it is extremely important to determine the crack path when evaluating G for ceramic joints.

In our analysis, the formulas for determining K_I and K_{II} for the homogeneous cracked disk specimen were obtained from the analysis of Atkinson et al. [9]. The formula for the homogeneous cracked bend bar was obtained from Reference [10].

4. RESULTS

4.1 Mechanical Test Results

Figure 4 is a Weibull plot of the apparent bend strength of zirconia/NCI joints. The average strength of the joint was 223 MPa compared with an average strength of 589 MPa for the as-received zirconia. We believe that an important reason for the strength degradation was the lower fracture toughness of the joint compared with the base zirconia. An additional strength reduction factor may have been an increased size of joint defects; partial evidence for a different flaw population compared with the base zirconia was the Weibull modulus of 5.4, which was smaller than the value of 13.4 measured for the base zirconia [3].

The cracked bend specimens provided an average G_c of 210 N/m. Results from the cracked disk specimens with θ=0 degree provided an average G_c of 236.8 N/m. These values were significantly less than G_{Ic} of 399 N/m for the as-received zirconia. One fracture toughness test was conducted where the crack was machined only in the reaction zone of the zirconia, to obtain its fracture toughness. This test provided a G_{Ic} of 237.5 N/m for the reaction zone, which was similar to that measured for the joints under nominally Mode I loading. This similarity in G_c values was in agreement with observations that a large fraction of fracture occurred through the reaction zone of the zirconia.

Figure 5 shows the variation of G_c with the mixity, ψ. The solid line is a secant-square least-squares best-fit to the data. The expression

$$G_c(\psi) = G_c(0)\mathrm{Sec}^2(\psi/\psi_o) \qquad (18)$$

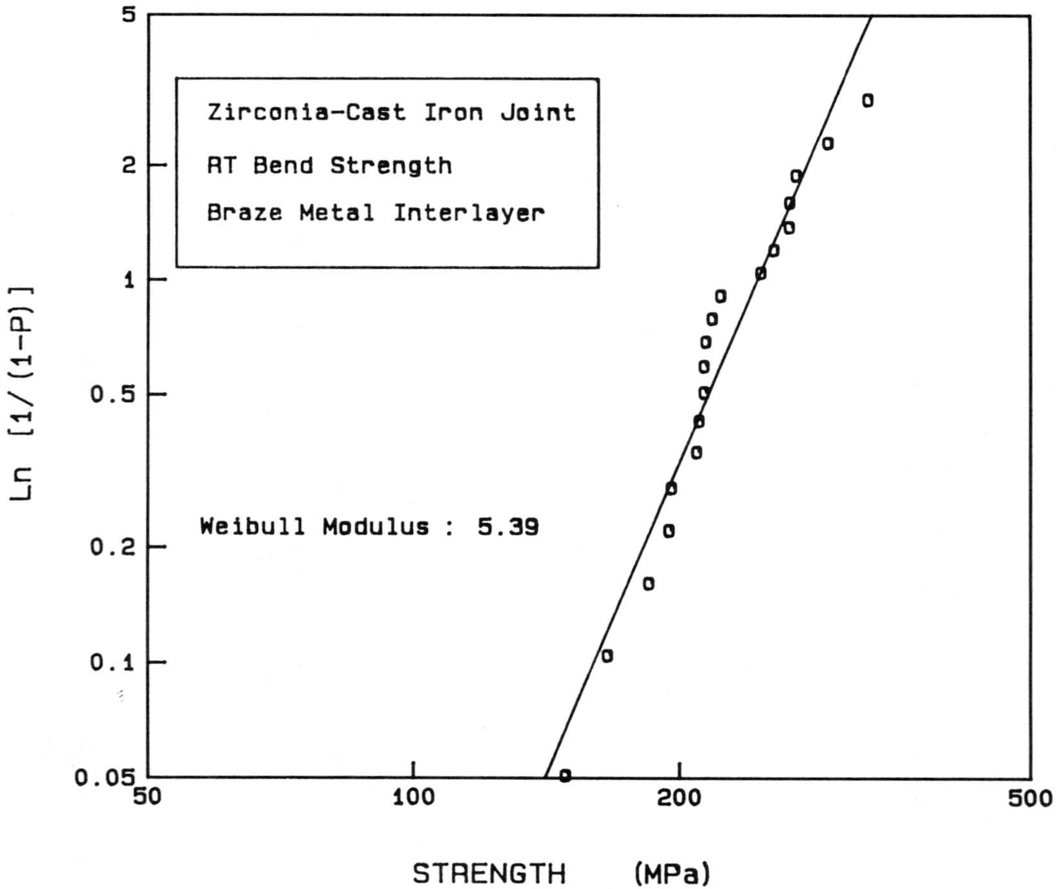

Figure 4. Weibull plot of the bend strength of zirconia/NCI joints

was used to fit the data, and was obtained from References [8,11]; the only modification was that a term ψ_o was introduced as a fitting parameter. Based on data available for monolithic ceramics, this fitting parameter has been found to vary between 1 and 2. In Figure 5, $G_c(0)$ was assumed to be 237.5 N/m, which corresponded to the G_{Ic} of the reaction zone. Thus, the only fitting parameter in equation (18) was ψ_o.

Figure 5 shows that up to about 40 degrees, there was very little change in the value of G_c. However, beyond that angle, there was a significant (almost three-fold) increase in G_c. This implies that the joint is significantly tougher under mixed-mode loading than under Mode I loading. Thus, in actual designs, it may be useful to alter the geometry of the joint such that some of the tensile loads are compensated for by shear loads, thereby taking advantage of increased G_c with increasing ψ.

Figures 6 and 7 are G_c versus ψ plots for zirconia/zirconia joints [3] and for monolithic alumina [12], respectively; the former was fabricated using a ceramic interlayer. These plots also illustrate the increasing nature of G_c with ψ, and suggest that a secant-square dependence may be quite universal.

4.2 Microstructure and Fractography

Figure 8 is a low magnification scanning electron micrograph (SEM) of a zirconia/NCI joint. The NCI is on the left of the figure, the zirconia on the right, and the central band corresponds to the inhomogeneous microstructure of the braze metal. Figures 9a and 9b are respectively high magnification BSE image and Ti-map of the braze-metal/zirconia interface. These figures illustrate that there was an almost complete segregation of Ti to the ceramic interface; we believe this was responsible for successful zirconia/NCI joints. The Ti segregation also was accompanied with depletion of oxygen from the zirconia substrate [13]; the reaction zone of the zirconia extended over a 0.4 mm zone, and was visible optically by virtue of the color being dark compared with the yellow color of the as-received zirconia.

Figure 10 is a fractograph of the NCI side of a NCI/zirconia joint; an unnotched bend bar was tested. EDS analysis showed that the flat featureless regions, such as 1, contained primarily Ti. On the other hand, the void containing projections, such as 2 in Figure 10, were pure zirconia. Figure 11 is a side view of the fracture, with the NCI on the left, followed on its right by the braze metal, then an empty cracked space, and finally zirconia on the right; the fracture halves are not

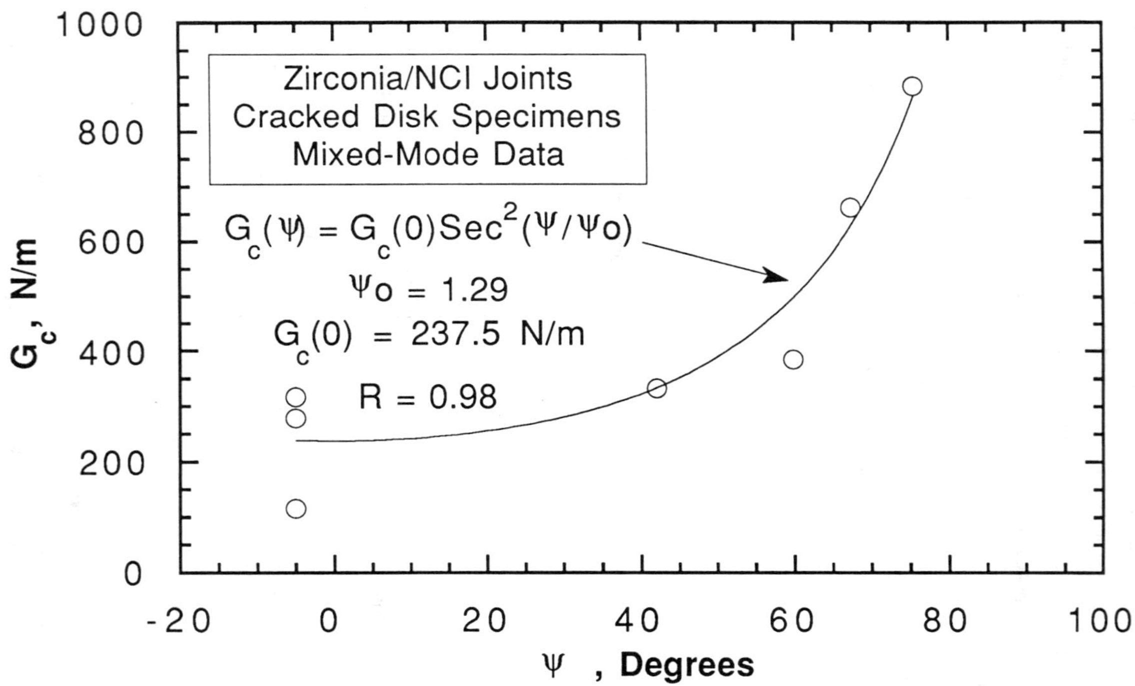

Figure 5. Mixed-mode fracture toughness data of zirconia/NCI joints plotted in the form of critical strain energy release rate, G_c, versus the mode mixity, ψ (=$\tan^{-1} K_2/K_1$), where K_1 and K_2 are the local Mode I and Mode II stress intensity factors, respectively.

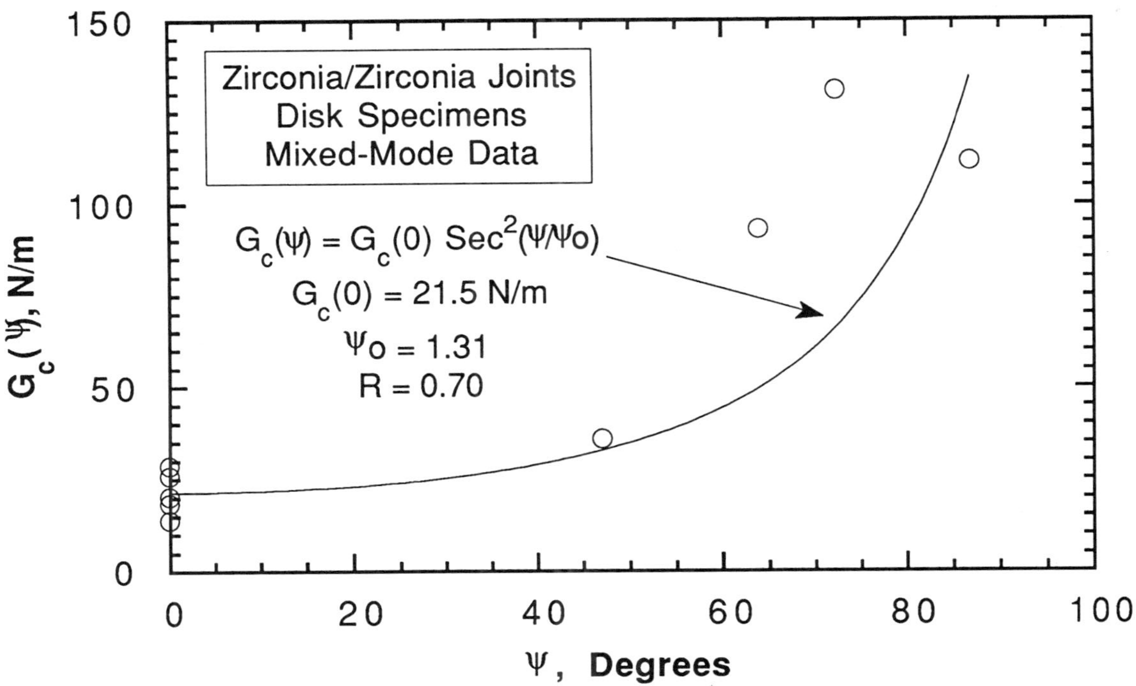

Figure 6. Mixed-mode fracture toughness data for zirconia/zirconia joints fabricated using a ceramic interlayer [3,4].

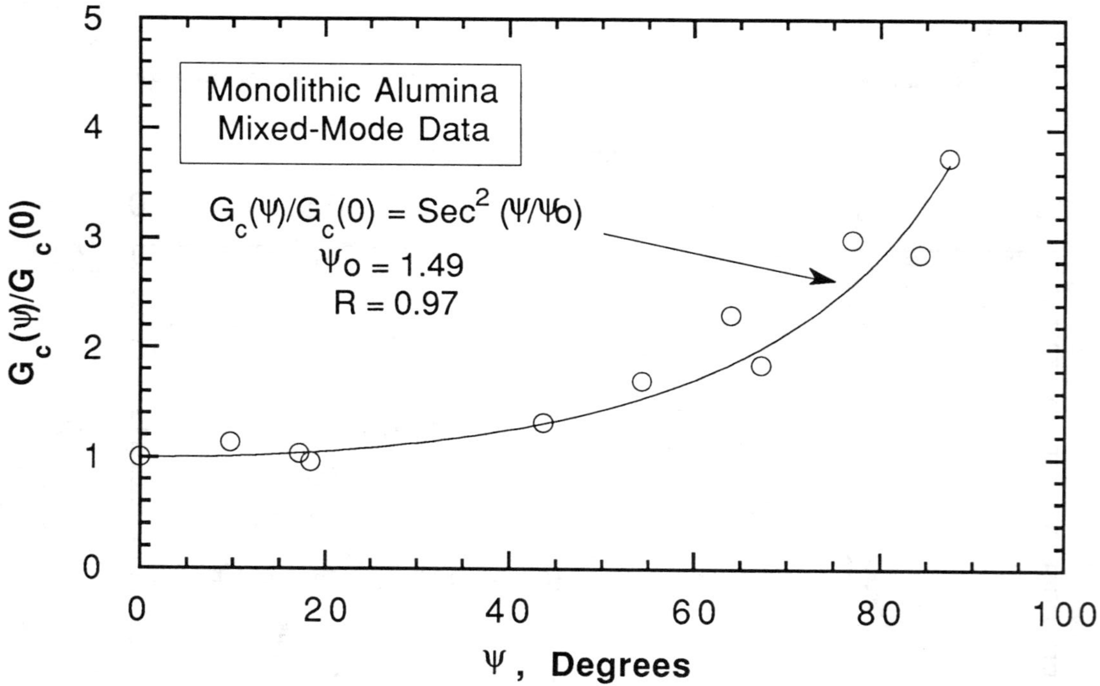

Figure 7. Mixed-mode fracture toughness data of monolithic alumina [12]. In this plot, the G_c has been normalized with respect to G_{Ic} (at ψ=0).

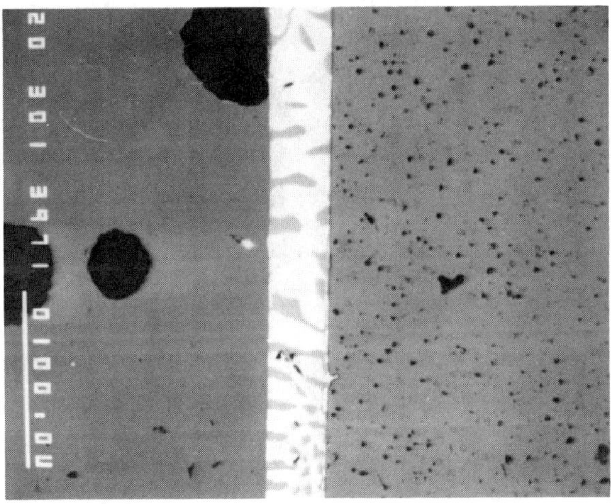

ure 8. Low magnification scanning electron micrograph of a zirconia/NCI joint. The NCI is on the left, the braze metal is at the middle, and the zirconia is on the right of the micrograph. The dark circular patches in the NCI correspond to the spheroidized graphite.

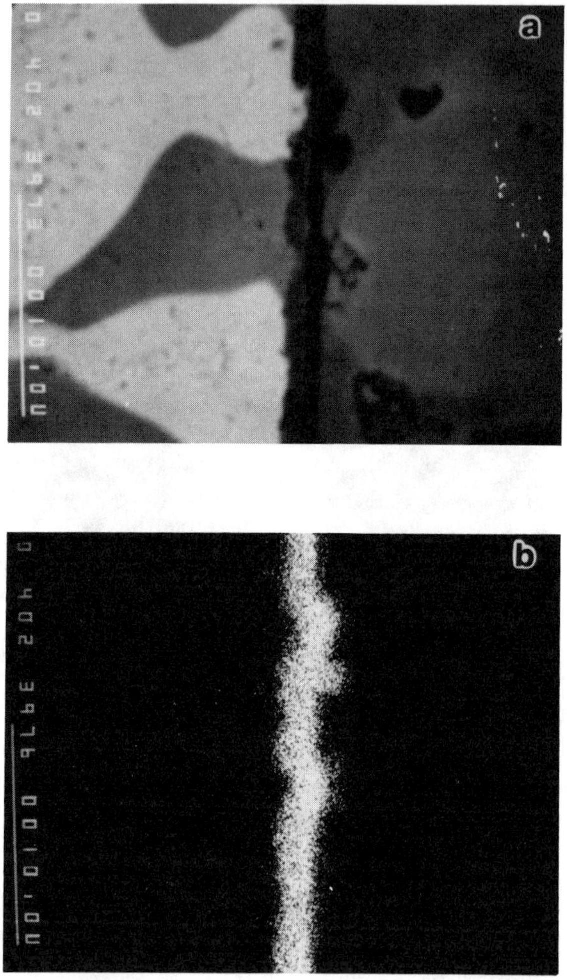

Figure 9. High magnification microprobe images of the joint, with the braze metal on the left and zirconia on the righ each micrograph. (a) is a back scattered electron (BSE) image, and (b) is a Ti-map of the same region. The differences in contrast in the braze metal arise because of inhomogeneous distribution of elements, nam Ag, Cu, In and Ti, with their accompanying differences in atomic number (Z); higher Z is associated with greater brightness.

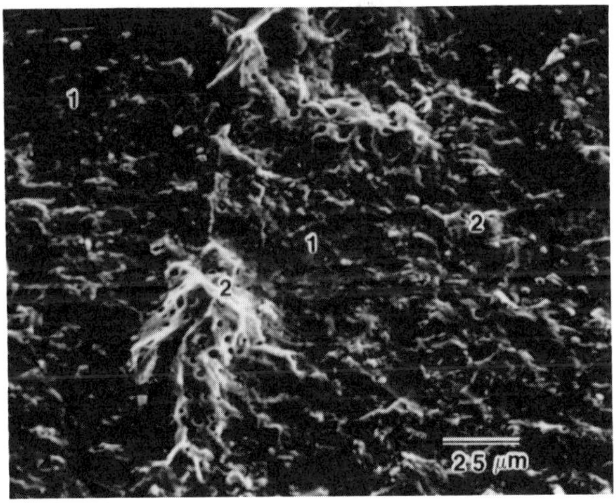

Figure 10. SEM image of the fracture surface of a zirconia/NCI joint, taken from the NCI side of the joint.

Figure 11. Side view of fracture of a zirconia/NCI joint; NCI is on the left, followed by the braze metal, then an empty cracked space, and finally zirconia on the right of the micrograph. Because of slight misalignment in the fractured pieces during mounting, the fracture halves in the micrograph do not correspond to mating surfaces.

mating surfaces. Figures 10 and 11 illustrate that the fracture path was oscillatory; i.e., meandering between the zirconia reaction zone and the Ti-rich braze/zirconia interface. We would like to comment that this fracture morphology is not just characteristic of zirconia/NCI joints. We have also observed this fracture path in other ceramic-metal systems, such as the alumina/titanium alloy system [14].

With increased mixity of loading ($\psi>0$), there was a gradual change in the fracture path and fracture morphology. On the zirconia side of the disk specimen, the crack deviated from the notch plane and propagated towards the loading point. This behavior was similar to what has been observed for monolithic ceramic specimens. The extent of crack growth along the bond-line prior to crack deviation (for zirconia side) depended on the loading angle, being approximately 6 mm for an angle of 10 degrees, and being less than 1 mm above 15 degrees. On the NCI side, however, the crack could not propagate into the NCI; rather the crack propagated all along the bond line. Figures 12a and 12b correspond to the NCI side and zirconia side, respectively, of the fracture surfaces of a disk specimen where the crack propagated through the bond line. The sample was loaded with the notch at an angle of 20 degrees with respect to the load line (ψ = 80.4 degrees). EDS analysis indicated that on the NCI side, the surface layer was rich in Ti and Cu, with Ti being the dominant element. However, contrary to Mode I fracture, hardly any zirconia was found on the fracture surface. The wavy structure seen in Figure 12a probably corresponds to shear deformation of the Ti-rich layer and of the braze metal beneath it. On the zirconia side, the primary elements detected were Zr and Ti and Cu. Here too, very little zirconia appeared to have been pulled out by the braze metal during fracture. Overall, it appeared that under shear loads there was significant plastic shear deformation of the braze metal, including the Ti-rich layers, and these resulted in shear failure of the interface without any significant accompanying failure of the reaction zone of the interlayer. Such fracture morphology differences between mixed-mode and primarily Mode I loading probably were responsible for elevating the G_c with increasing ψ.

5. DISCUSSION

Zirconia/NCI joints were successfully fabricated using an active-filler metal approach. These joints were characterized using two significantly different geometries, namely, the bend bar and the disk specimen. The fracture toughness of joints was characterized in terms of the energy release rate, G, as a function of mixity, ψ. As indicated in Reference [4], this approach is more appropriate for joints rather than K, because an additional length parameter enters into the stress equations.

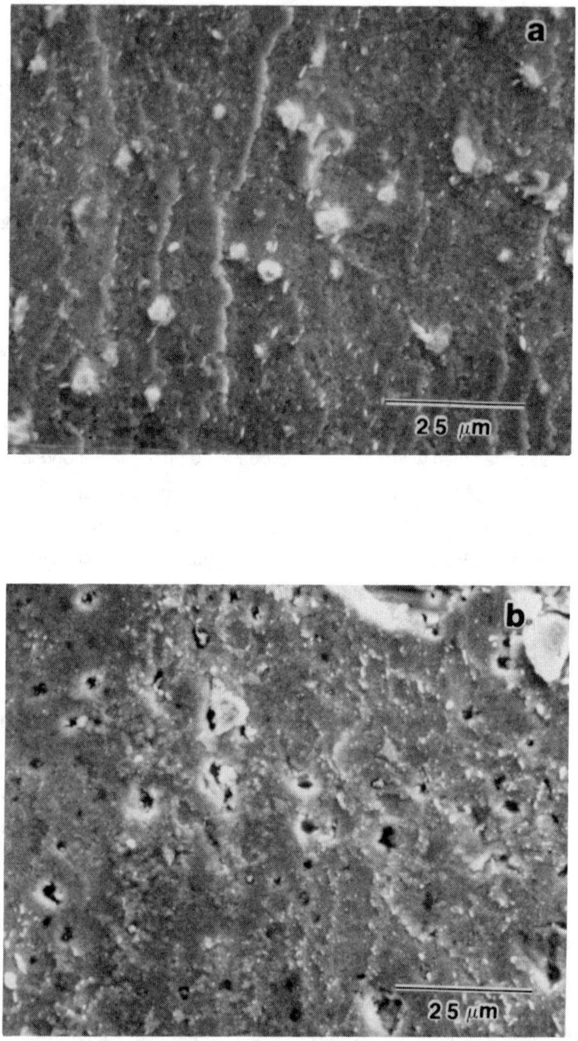

Figure 12. Fracture surfaces of a zirconia/NCI joint that was fractured at a mixity value of 80.4 degrees; (a) and (b) correspond to the NCI half and zirconia half of the joint, respectively.

Because of the close match between the modulus of zirconia and NCI, simple asymptotic solutions [8] (equations 10 and 11) were used to determine G as a function of ψ. The approach that was taken was to obtain an average modulus for the substrate, and treat the fracture problem as an interface crack with a thin interlayer interposed between two materials of identical elastic properties. FEM analysis showed that the asymptotic solution was in excellent agreement with the FEM results [3,4].

For nominally Mode I loading, the G_c was determined to be approximately 210 N/m using the bend bar geometry, and approximately 237 N/m using the cracked disk geometry. This difference in G_c values is small considering the inherent scatter of most fracture toughness data. Thus, the results may be viewed as establishing the geometry independence of a fracture mechanics approach for predicting fast fracture of cracked ceramic-metal joints.

It was found that beyond ψ=40 degrees, there was a sharp increase in G as a function of ψ. This characteristic suggests that a possible design for a butt-joined assembly is not to have the joint perpendicular to the loading direction. Rather, provided there are no processing and geometrical constraints, a more efficient design would be to have a joint at an angle to the loading direction, taking advantage of increased G with increased ψ. Also, from a joining viewpoint, a joining technique that provided earlier and faster increase of G with ψ would be beneficial. This characteristic of mixed-mode failure has to our knowledge not been exploited in the design of ceramic-metal joints.

The microstructure of joints showed that Ti was primarily responsible for successful bonding. However, oxygen depletion of the adjacent zirconia adversely affected the properties, and efforts should be made to reduce the extent of degradation. Fractography showed that failure occurred partly through the reaction zone of the zirconia and partly through the Ti-rich zone of the braze metal/zirconia interface. This provided guidance in the analysis procedure. In fact, observation of fracture path in ceramic-metal joints is of critical importance, because the fracture characterizing parameters are sensitive to the failure location. With increased mixity of loading, shear deformation of the Ti and Cu layers were observed. These may have been responsible for increased G with ψ. However, additional tests and further detailed examination is necessary.

A convenient method applicable to the design of joints is possible if the results are presented in the framework of a failure assessment diagram [3,4]. Shown in Figure 13 for an example case, the area enclosed by the axes (or a part of this area) may be taken to be the design space. The assessment vector is calculated corresponding to a reference design load, P_A. For example, P_A

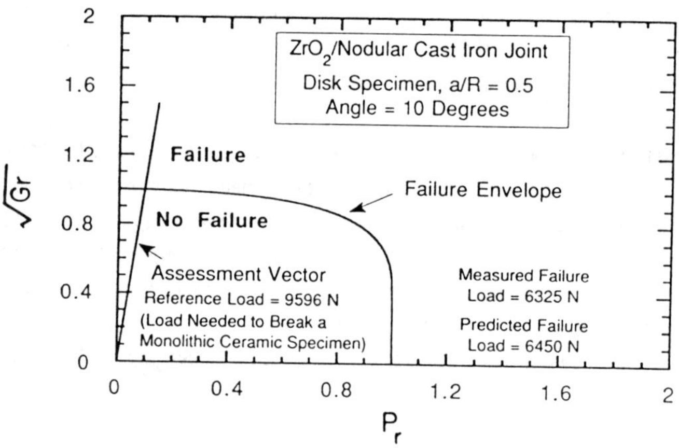

Figure 13. Assessment Diagram for a zirconia/NCI disk specimen

may be taken as the smaller of the failure loads of identical monolithic specimens of the substrate materials. If the assessment vector just touches the failure curve, the joint will be able transfer as much load as a specimen of the weaker substrate material. By varying interlayer properties, one may try to minimize the assessment vector length for design improvement. A more detailed description of this joint assessment and design model approach is contained in Reference [3].

6. CONCLUSIONS

Zirconia-NCI joints were successfully fabricated using the active filler metal approach. Ti in the braze metal primarily was responsible for good bonding.

The strain energy release rate, G, was used for characterizing the fracture toughness of joints. Experiments were performed using two significantly different geometries: the cracked 4-point bend bar and the cracked diametrically compressed disk specimen. For nominally Mode I loading, these two geometries provided G_c values that were in good agreement with one another (within 12 percent), thus establishing geometry independence of a fracture mechanics based approach.

Fracture was found to occur partly through the zirconia and partly through the braze metal/zirconia interface. Because the evaluation of G is strongly dependent on failure location, it is extremely important to determine fracture path in any ceramic-metal fracture toughness evaluation.

The strength and toughness of joints were found to be lower than those of the base zirconia. The degradation of zirconia in the reaction zone is believed to be primarily responsible for the reduced properties.

G_c was found to be a strongly increasing function of mode mixity. It is suggested that this characteristic be utilized in the design of components containing ceramic-metal joints.

7. ACKNOWLEDGMENTS

The authors wish to thank Drs. M. Santella and D.R. Johnson of Oak Ridge National Laboratories for supporting the present work under subcontract 86X-SB046C. The contributions of Prof. J. Cawley (The Ohio State University) and Dr. D. Hauser (Edison Welding Institute) in the fabrication of joints are gratefully acknowledged. We also acknowledge with thanks the useful discussions with Drs. A.T. Hopper and A.R. Rosenfield of Battelle.

8. REFERENCES

1. M. Turwitt, G. Elssner, and G. Petzow, J. de Physique, Vol.46, pp.C4-123 (1985)

2. H.F. Fischmeister, G. Elssner, B. Gibbesch, and W. Mader, Proceedings of MRS International Meeting on Advanced Materials, Tokyo, paper F5.5 (1988)

3. J. Ahmad, B.S. Majumdar, A.R. Rosenfield, S.L. Swartz, J. Cawley, E.Park, D. Hauser and A.T. Hopper," Analytical and Experimental Evaluation of Joining Ceramic Oxides to Ceramic Oxides and Ceramic Oxides to Metal for Advanced Heat Engines Applications", Battelle Memorial Institute Report to Martin Marietta Energy Systems, October (1990)

4. J.Ahmad and B.S. Majumdar,"An Engineering Fracture Mechanics Analysis of Metal-Ceramic and Ceramic-Ceramic Joints", Proceedings of ASTM 22nd Symposium on Fracture Mechanics (1990)

5. J.R. Rice,"Elastic Fracture Mechanics Concepts for Interfacial Cracks", Journal of Applied Mechanics, Vol. 55, pp.98-103 (1988)

6. M. Cominou,"The Interface Crack", Journal of Applied Mechanics, Vol 44, pp.631-636 (1977)

7. C.F. Shih and R.J. Asaro,"Elastic-Plastic Analysis of Cracks on Bimaterial Interfaces: Part 1-Small Scale Yielding", Journal of Applied Mechanics, Vol. 55, pp.299-316 (1988)

8. J.W. Hutchinson,"Mixed Mode Fracture Mechanics of Interfaces", Proceedings of Acta/Scripta Metallurgical Conference on Bonding, Structure and Mechanical Properties of Metal/Ceramic Interfaces (1989)

9. C. Atkinson, R.E. Smelser, and J. Sanchez, "Combined Mode Fracture via the Cracked Brazillian Disk Test", Int. J. of Fracture, Vol. 18, No. 4, pp.279-291 (1982)

10. D. Broek, Elementary Engineering Fracture Mechanics, Sitjoff and Nordhoff International Publishers, Holland (1978)

11. A.G. Evans and J.W. Hutchinson,"Effects of Non-Planarity of the Mixed-Mode Fracture Resistance of Bimaterial Interfaces", Acta Metallurgica, Vol. 37, p.909 (1989)

12. D. Singh and D.K. Shetty,"Fracture Toughness of Polycrystalline Ceramics in Combined Mode I and Mode II Loading", J. Amer. Ceramic Society, Vol.72, pp.78-84 (1989)

13. J.P. Hammond, S.A. David, and M.L. Santella,"Brazing Ceramic Oxides to Metals at Low Temperatures", Welding Journal, Vol. 67, No. 10, pp.227-232 (1988)

14. B.S. Majumdar and K. Ramesh, unpublished research

LIST OF FIGURES

Figure 1. Specimen geometries used for determining fracture toughness of joints and base materials. The left and right figures correspond to the straight-notched cracked bend-bar and the chevron-notched diametrally compressed disk specimen, respectively.

Figure 2. Residual stress (σ_{xx}) profiles for a zirconia/NCI bend bar (uncracked specimen). The y-axis corresponds to the depth of the beam and the x-axis corresponds to the length of the beam.

Figure 3. The fracture mechanics problem analyzed in this investigation. The crack location was based on experimental observation of fracture path of an uncracked specimen.

Figure 4. Weibull plot of the bend strength of zirconia/NCI joints.

Figure 5. Mixed-mode fracture toughness data of zirconia/NCI joints plotted in the form of critical strain energy release rate, G_c, versus the mode mixity, ψ (=$\tan^{-1} K_2/K_1$), where K_1 and K_2 are the local Mode I and Mode II stress intensity factors, respectively.

Figure 6. Mixed-mode fracture toughness data for zirconia/zirconia joints fabricated using a ceramic interlayer [3,4].

Figure 7. Mixed-mode fracture toughness data of monolithic alumina [12]. In this plot, the G_c has been normalized with respect to G_{Ic} (at ψ=0).

Figure 8. Low magnification scanning electron micrograph of a zirconia/NCI joint. The NCI is on the left, the braze metal is at the middle, and the zirconia is on the right of the micrograph. The dark circular patches in the NCI correspond to the spheroidized graphite.

Figure 9. High magnification microprobe images of the joint, with the braze metal on the left and zirconia on the right of each micrograph. (a) is a back scattered electron (BSE) image, and (b) is a Ti-map of the same region. The differences in contrast in the braze metal arise because of inhomogeneous distribution of elements, namely Ag, Cu, In and Ti, with their accompanying differences in atomic number (Z); higher Z is associated with greater brightness.

Figure 10. SEM image of the fracture surface of a zirconia/NCI joint, taken from the NCI side of the joint.

Figure 11. Side view of fracture of a zirconia/NCI joint; NCI is on the left, followed by the braze metal, then an empty cracked space, and finally zirconia on the right of the micrograph. Because of slight misalignment in the fractured pieces during mounting, the fracture halves in the micrograph do not correspond to mating surfaces.

Figure 12. Fracture surfaces of a zirconia/NCI joint that was fractured at a mixity value of 80.4 degrees; (a) and (b) correspond to the NCI half and zirconia half of the joint, respectively.

Figure 13. Assessment Diagram for a zirconia/NCI disk specimen.

Joining Processes

PROGRESS IN CERAMIC-METAL JOINING AND METALLIZATION:

AN OVERVIEW OF TECHNIQUES AND RECENT ADVANCES

Victor A. Greenhut

Department of Ceramics
College of Engineering

Rutgers - The State University of New Jersey
Piscataway, NJ 08855-0909

Abstract

An overview is provided of the various techniques available for joining ceramics and metals and for metallization of ceramics:

1. Sintered Metal Powder Process [SMPP]
2. Metal Powder - Glass Frit
3. Active Metal Brazing
4. Glass Bonding / Non-metallic Fusion
5. Gas Metal Eutectic
6. Pressed Diffusion
7. Vapor Phase Metallizing
8. Liquid Phase Metallizing
9. Electroforming
10. Graded Powder Process

The procedure for each process is outlined and important considerations are discussed. Recent improvements in joining/metallization methods are reviewed for each of the bonding methods.

Metal - Ceramic Joining
Edited by P. Kumar and V. A. Greenhut
The Minerals, Metals & Materials Society, 1991

Introduction

Metals and ceramics are frequently joined with one of the components as a liquid phase and the other solid. Deposition from a vapor or solid-solid joining may also be used. With intimate physical contact, bonding may occur by surface or interfacial interaction. Solution and reduction/oxidation interactions may proceed which promote good adherent bonds. In technologies where a liquid phase is involved it is a usual requirement that wetting by the liquid phase occur. When obvious melting occurs the liquid phase is obvious as, but sometimes the liquid phase is a thin layer or a microscopic constituent. Often interactions between the materials themselves and with the environment will alter the wetting response with time. While a wetting response is usually sought in order to create a satisfactory bond, it is not entirely necessary and is not a guarantee of strong bonding.(1-6)

Stresses between joined components may also affect the bond importantly, either deteriorating bond quality, causing components to delaminate from seals or resulting in interfacial failure. Such stresses may arise because of differences in the elastic modulus under an applied or residual stress. Differences in thermal expansion or phase transformations can give rise to stresses which exceed the bond strength. Thermal cycling may introduce cyclic fatigue due to this expansion difference.

A common feature of the more successful techniques for joining ceramics and metals is a transition between the metallic bond and the covalent/ionic bond of the ceramic. In glass sealing to metals (nonmetallic fusion) it may be best to have a transition in the activity of metallic specie in the glass near the bond interface (5). A mixed interfacial glass and metal structure is formed in the sintered metal powder process which forms a transition to metallic structure for conventional metal brazing technologies (7). It is common for metal to be preoxidized in glass-to-metal joining and sealing (3,7,8). The oxide layer adheres to the metal and may lower the interfacial energy of the surface thereby promoting wetting by providing a covalent/ionic bond more compatible with the glass. It is quite common to employ a "forming" gas which will tend to oxidize the metallic component or reduce the ceramic or glass partially at the surface. Such surface modification can promote wetting and significantly enhance bonding. Indeed, atmosphere control is often vital to successful bonding between metals and ceramics. All of the above techniques suggest that a transition between metallic and covalent/ionic bonding is an important consideration for successful wetting and adherence.

Bonding Mechanisms

A convenient system for considering and promoting bonding of ceramics and metals is division into: physical, mechanical and chemical bonding. Generally, all three mechanisms are found in well bonded systems and it is difficult to separate the contribution of each. Enhancing the mechanisms in preparing the material for bonding and in the joining process is important in order to engineer a strong bond.

Physical bonding is the effect which occurs when two perfectly flat surfaces are brought together to atomic interaction distances. Local atomic rearrangement results in adhesion. The difference of energy between the specific surface energy of the two materials is the work of adhesion. The work of adhesion can yield a theoretical breaking stress similar to the strength of the metal or ceramic (4,9). The actual failure stress is often several orders of magnitude smaller as the result of flaws in the bond. Physical bonding provides a guideline to the selection of systems which will bond well.

Mechanical bonding refers to the interlocking structure of rough surfaces providing interlocking of microstructure to provide tensile strength and frictional strengthening in the case of shear. During the bonding process liquid metal or glass can flow into cavities and asperities, a ductile metal can conform to a rough ceramic surface, or a vapor may deposit in surface asperities. The surface may be roughened by acid or base chemical attack, electrolytic corrosion, scouring, grinding or grit/sand blasting to enhance the mechanical bonding. In grit blasting passes in multiple directions at an inclined angle aids in producing undercuts which enhance the bond. The effectiveness of different surface roughening methods is dependent upon their optimum application as well as the specific method and materials used. Chemical interaction between the materials being bonded may also lead to mechanical bonding. The increase in surface area of the rough surface also may provide an increase in physical bonding.

The degree of mechanical bonding is often modeled upon the interface ratio or undercut density. The former is the ratio of actual bonded surface to that of a flat smooth interface and is often taken on a cross section sample. The latter measures the number of re-entrant features on a cross section per linear distance. An increase in either number usually enhances the bond, but undercut density tends to correlate better with tensile adherence while either measure may correlate with shear failure strength.

Chemical reactions may change the nature of the bond interface providing a transition zone by dissolution, interdiffusion, redox, and precipitation. As the bond between ceramics and metals is fundamentally different, a theory of bonding first developed for glass-to-metal bonds may have general applicability. The model hypothesizes that the glass (ceramic more generally) at the interface must be saturated with the lowest oxidation state oxide of the metal being bonded (5, 10). There is an idealized molecular monolayer of metal oxide at the interface providing a continuity of chemical activity (bond energy) and structure between metal and ceramic. Incorporation of the metallic species may either occur by diffusion or reaction during bonding or by prior processing. Incorporation of metallic specie in the ceramic or glass can promote bonding. This is often termed the chemical or oxide layer theory.

In glass-to-metal bonding electrolytic theory may operate. Adherence promoting metal oxides dissolved in a glass are reduced during firing to a metallic state by the base metal. Alternatively a flash of the metal may be deposited on the base metal. An example is the use of cobalt either in the base coat or as a flash in the porcelain enameling of cast iron and steel. A local galvanic cell is formed which strongly corrodes the base metal. The metal going into solution provides chemical bonding through the continuity of activity discussed above. The rough surface with anchoring etch pits provides mechanical adhesion. Supersaturated metal may be precipitated during this process or upon cooling, resulting in dendrites extending from the base metal. This gives a further mechanical anchor into the glass termed the dendritic theory.

Other chemical reactions may produce a transition structure that varies from metallic to ceramic. Often metals are preoxidized to give an oxide that will be wet by and react with glass itself or glass within a ceramic saturating the interfacial region. An adherent oxide on the base metal may provide a ceramic surface for bonding to a glass. A physical mixture of ceramic and metal such as that in the sintered metal powder process (see below) may provide a transition structure between the metal and ceramic or glass. Interfacial phases which form may have properties different from either metal or ceramic. (1- 10)

In practice complex combinations of the various mechanisms actually occur. Thus, features of physical, mechanical and chemical adhesion can all be identified in actual practice, but their individual contributions may be difficult to determine. (11,12)

Metallization/Joining Methods

Ceramics and metals can be joined with fasteners or adhesives, but when a durable, hermetic join is required capable of extended service in a variety of environments and temperatures a direct bond between metal and ceramic is desirable. Such bonds require a preparation step which will provide a transition structure from ceramic covalent/ionic bonding to the metallic bond. This will assure wetting and chemical adherence between ceramic and metallic parts. Often these methods are characterized by the way in which bonding or sealing materials are applied, ie : Liquid phase, solid phase or vapor phase joining.

Comprehensive reviews of ceramic metallization and joining can be found in references 3, 12, 13 and 14. Surface preparation of the ceramic is important. Often the surface chemistry and structure of a fired ceramic is quite different from the bulk. The machining, grit blasting, polishing, or chemical etching of the ceramic surface may substantially decrease or increase adhesion of a metallizing layer. If the ceramic surface is processed or handled after firing cleaning and removal of adsorbed water or hydrocarbons may be needed. Chemical cleaning, acid etching, or grit blasting may be preliminary steps. For ultimate cleanliness the ceramic is baked out briefly at elevated temperature (800 to 1500 deg. C). As-fired ceramic may need no cleaning to produce a satisfactory bond.

1. Sintered Metal Powder Process

The sintered metal powder process is often referred to as "SMPP" or the "moly-manganese paste process". It uses a paste composed of molybdenum and manganese metals together with their oxides suspended in an organic vehicle. Compositions containing tungsten, iron, tantalum, rhenium, and titanium with their oxides and in combination with molybdenum and manganese can also be used. Some "moly-manganese" pastes contain neither molybdenum or manganese. A typical composition is:

60 g	-	molybdenum powder	(0.2 um)
15 g	-	manganese powder	(0.2 um)
0.2 g	-	nitrocellulose	
30 mL	-	butyl acetate.	

The paste is applied by brushing, spraying, transfer tape, roller application, silk screening, and dipping. The particle size organic binder and vehicle are adjusted to provide suitable rheology for the particular application method. A layer thickness of about 20 um is typically applied to a clean as-fired, machined or ground ceramic surface. The higher glassy phase content near the surface of many as-fired ceramics has been found to enhance bonding The SMPP process has been used successfully to metallize a wide variety of oxide, carbide and nitride ceramics and ceramic composites Ceramics containing little or no glassy silicate phase require the incorporation of such material into the paste composition. For example to bond to a 99% alumina or sapphire, the composition above may be mixed with an equal amount of powder composed of :

13 Al_2O_3, 52 MnO, 35 SiO_2 or
41 Al_2O_3, 54 CaO, 5 MgO.

After application of the paste, drying is performed in air often using heat lamps. Firing is done in a hydrogen atmosphere (dew point approximately 25 deg. C.) at 1200 to 1600 deg. C for about a half hour. Titanium and manganese additions or surface vapor deposited coatings may enhance bonding and lower the firing temperature a few hundred degrees.

The structure of the coating provides a transition from a mixed glassy and crystalline structure which "migrates" from the ceramic and the paste to ceramic interface to a metallic structure formed by hydrogen reduction at the surface of the metallization (1, 2, 7, 15). The structure is that of a liquid phase sintered product consisting of a glassy phase with metallic and oxide ceramic particles. Near the ceramic interface the particles are generally more ceramic in nature while at the surface metal particles predominate. This gives a bridging structure between ceramic and metallic bonding.

It is difficult to braze this metallized surface directly because most brazes do not wet the metallization. It is more common to electroless or electroplate nickel onto the surface to a layer thickness at least 5 to 8 um. Copper or other metals can also be plated. In some cases a metal powder is applied and allowed to diffusion bond to the surface in a reducing atmosphere. The nickel coated surface can be brazed employing conventional metal joining brazing technologies. In electronic applications the nickel or copper coating may be a conductor or contact in itself or may be soldered or wire bonded.

The SMPP process requires a number of steps and unit operation. The procedures are relatively easy to engineer and can yield reliable, consistent results. The bond is amongst the strongest achieved. It is the most common metallization and ceramic-to-metal joining method in areas other than microcircuitry. It has the advantage of allowing final joining of a metal to the metallized ceramic by conventional brazing methods. This makes it amenable to mass production, as well as the joining of large or complex shapes and structures.

Recent advances have been made bonding to non-oxide ceramics such as silicon carbide, aluminum nitride and silicon nitride. In many cases preoxidation of the surface is a necessary condition either for a strong or reliable bond. The composition of the SMPP pastes are proprietary, but based upon conventional formulations.

Increased attention has been drawn to the ceramic surface to be metallized. Surface finish may play an important role, but there is evidence that chemical differences between an as-fired ceramic and a polished material may have a greater significance. Vapor deposition of active metal may enhance the bond. (16)

2. Metal Powder - Glass Frit Method

For the metallization of hybrid microelectronic circuitry and to some extent for some other ceramic-to-metal joins the metal powder-glass frit method is used. In the precious metal powder-glass frit method a noble metal alloy based on platinum, palladium, gold, silver and/or iridium in finely divided form is mixed in a glass frit and suspended in an organic binder and vehicle. Control of particle size, composition and organics has yielded a range of rheological properties for various application methods. The pattern is usually silk screen printed onto an electronic substrate material such as alumina or a multilayer capacitor material such as barium titanate. One advantage of the precious metal systems is that they permit air firing. The glass composition is selected to provide low viscosity for good flow at the desired bonding temperature. A bonding glassy phase forms at the ceramic-metal interface to provide adherence. The surface is metal rich and may be used directly as a conductor, be plated, soldered or brazed without further surface preparation. The transition from a glass bonded ceramic structure at the ceramic interface to a metallic structure at the surface is akin to the SMPP process and may also be regarded as a liquid phase bonding technology.

The as-fired surface of the ceramic may show better adhesion than a polished or etched surface. This is either because there is more glassy phase at the ceramic surface or because of differences in the grain microstructure. A clean, dry substrate is used and an as-fired alumina substrate may show better adhesion because of the glass content or microstructure at the surface. Active metals and rare earth metals may be included in the composition in order to enhance bonding. Alternatively active constituents may be vapor deposited on the ceramic before bonding (16). The incorporation of more active metals in the composition (titanium, zirconium, nickel, copper) may enhance the formation of mixed metal oxides at the ceramic-metal interface. These as well as glassy devitrifying phases may improve adherence (17). Incorporation of more active metal may require firing under inert or reducing conditions. Rare earth elements may be first vapor deposited on the surface in order to enhance adhesion.

Copper and nickel can be used in much the same way at a lower material cost than the precious metals. The glass frit must be a composition stable at low oxygen partial pressures (Al_2O_3, B_2O_3, CaO, MgO, SiO_2, ZnO) since the metal particles must be fired in a nitrogen atmosphere with a low oxygen partial pressure (10 ppm). Controlled oxidation of the metal is required in order to insure wetting of the metal particles by the glass. (17,18)

3. Active Metal Process

The active metal process involves the use of an active metal constituent such as titanium, zirconium or hafnium in a braze metal composition. This causes conventional brazing alloys to wet and adhere to a ceramic. The active metal may be alloyed with copper or nickel in order to reduce its reactivity. Brazing is usually performed at carefully controlled temperatures from 800 to 1200 deg. C in a vacuum or inert atmosphere. The braze metal may be applied as a powder, wire or shim directly to the ceramic. Because the metal part must be attached in the metallizing operation, joining by the active metal process has geometrical and size limitations. A hydride of the active metal may be applied as a separate layer on the ceramic and a more conventional braze used (19).

Active metal is transported in the liquid braze to the ceramic interface and interacts forming a transition microstructure (20). Wetting can be so enhanced by the use of active metal that great care must be taken to prevent braze metal from flowing into unwanted areas. Bonds are amongst the strongest observed for ceramic-to-metal joins. The active metal cannot be used at elevated temperature, because the reactions will continue with the ceramic and adherence will deteriorate. For the same reason process control is critical to obtaining reproducible, high quality bonds. The application of a thin layer of active metal by vapor phase deposition techniques has been found to improve bonding. Recently new compositions have been developed to bond non-oxide structural ceramics for elevated temperature use. (16)

4. Glass Bonding / Non-metallic Fusion

In glass-to-metal sealing the materials are usually selected both for required engineering properties and also to provide thermal expansions which match reasonably over the temperature range of fabrication and use. After forming and metal cutting operations the metal is usually cleaned chemically in caustic and washed. It may then undergo acid pickling, neutralization and washing followed by thorough degreasing. A plating or metal flash may be introduced to promote bonding, in which case the sequence above is repeated. The metal then is commonly decarburized and/or hydrogen baked out both to clean the surface an diffuse out specie which will produce gas. Other products may also need to be removed, such as nitrogen in the case of titanium and its alloys. Frequently the metal then undergoes a controlled oxidation step. This creates a tightly adherent layer of oxide which must remain adherent during formation of the glass-to-meal bond. In some cases the adherence promoting oxide is produced in the same step as the glass sealing by appropriate atmosphere control and controlled viscous flow of the glass. The adherence oxide is usually wet by the glass and dissolves partially or wholly in the glass providing a chemical glass-to-metal bond (15). References 3,8,11 can be consulted for reviews of this subject.

Tungsten, molybdenum, titanium, tantalum, stainless steels, nickel, nickel-iron, F15-type, platinum, copper, gold, copper clad nickel and iron-nickel Dumet alloys have been used in common engineering applications. These metals have been coupled with a large range of borosilicate and lead silicate glasses. Aluminosilicate, lithium, barium, and soda-lime-silica glasses have been used to a lesser extent. Borosilicate glasses are most commonly used in glass-to-metal sealing, but lead silicate glasses are more common for bonding to copper, copper clad alloys, and iron-chromium alloys such as stainless steels. Barium is being substituted for lead based glasses recently in production applications where lead vapor may represent a potential hazard.

The glass is usually applied as a frit, paste, bead or solid preform. The glass may be heated to the required viscosity range by electrical, radio-frequency or flame heating either with the glass already applied to the preoxidized metal surface or presoftened and then pressed into contact with the metal. Graphite is frequently used to align or press parts together because it shows little bonding with the glass. Stopoffs, such as milk of magnesia, may be used to guide glass flow or prevent its flow into unwanted ares. It is common for the joining process to be conducted in a protective or special forming gas atmosphere to promote wetting and adhesion. Rapid cooling is avoided or an annealing step interposed to avoid the development of residual stresses due to rapid cooling.

ASTM F15 alloy, Fe- 29Ni -17Co, is a common glass sealing alloy because it provides very good expansion matching with borosilicate glasses. Commonly referred to by the Carpenter Technologies trade name "Kovar", it also can be oxidized to provide a controlled highly adherent oxide which is wet well by the glass and creates a strong bond. Oxidation along near surface grain boundaries of the alloy gives mechanical anchoring even without ostensible surface roughness. F15 alloy is used in microelectronic packaging, electrical feedthroughs, high frequency devices, windows and ports, and other diverse applications. It provides a common and convenient example of the procedure for glass-to-metal sealing.

After forming, the Kovar alloy is alkaline degreased, water washed, chemically cleaned, water washed and freon degreased with careful attention to the quality of final degreasing steps. If a shaped glass preform is used or if it has protective sizing (starch) it may be washed and cleaned with hydrofluoric acid followed by water washes and possibly freon or organic rinses. The metal then is subjected to a bake out and preoxidation treatment, while it is rare but possibly beneficial for the glass to be baked out. The Kovar is decarburized in hydrogen, hydrogen/nitrogen, or a dissociated ammonia atmosphere with controlled water level at 900 to 1100 deg. C. The Bake out temperature should be higher than the final sealing temperature. A typical cycle is 10 to 15 minutes at 1100 deg. C in a 1 vol% water, 10 vol% in hydrogen, nitrogen atmosphere. Some manufacturers repeat the procedure. The alloy should not be stored at this or subsequent steps as oxidation, contamination, and adsorbed water may interfere with the quality of bonding.

The F15 alloy is next oxidized in a controlled way. Kovar is sufficiently tolerant so that the alloy can be oxidized in air and natural gas, but bond quality will be variable chiefly because of variations in relative humidity and water content of the gas. Pure oxygen gives a thick nonadherent, flaky oxide which is unsatisfactory for bonding. When produced in pure oxygen the oxide does not penetrate along grain boundaries providing anchoring points. A high quality oxide structure with suitable thickness and grain boundary penetration can be accomplished with a 10 minute treatment at 800 to 1050 deg. C in a gas composed of 1 vol% water, 0.4 vol% hydrogen, nitrogen.

The preoxidized metal is now ready for glass sealing. The glass frit, bead, paste or preform is applied. The atmosphere may be like that of the preoxidation step or natural gas may be used with somewhat greater variability in adherence. Sealing is done at about 1000 deg C. Glass viscosity, wetting response, partgeometry, part loading, and part geometry determine the time and temperature used in the sealing step (21). A combined, single step procedure has been developed which eliminates the decarburization and preoxidation steps (22, 23).

The nonmetallic fusion process, the use of a glass to bond ceramic and metal, is similar to the technology of glass sealing. A glass frit is heated to its working range and is used to bond a metal and ceramic component. This liquid phase method has technological requirements similar to glass-to-metal bonding. Bonding to the ceramic is usually rather straightforward as the glasses generally wet and bond well. Thermal expansion, residual stresses and the formation of undesirable ceramic phases does need to be considered.

The glass layer is purposely kept thin. It must suffice for bonding to the metal but should be incorporated into the ceramic structure to the greatest extent possible. A residual thick glass layer may be a zone of structural weakness. Incorporation of the glass into the ceramic structure may also allow the joint to be used at temperatures higher than the softening or fabrication temperature of the bond.

An alternative is to use a bonding glass that devitrifies to form a glass-ceramic. This has the advantage of greater strength, a wider range of thermal expansions for matching purposes and better chemical durability than sealing glasses. An example is a lithium phosphate nucleated lithium silicate glass-ceramic which matches and seals well to Inconel and Hasteloy alloys (16).

A typical application of conventional nonmetallic fusion is the use of manganese aluminosilicate glasses which may be applied to the joint interface as a preform of suspension. The assembly is heated in inert gas at 1200 to 1500 deg. C for about 15 minutes and then cooled rapidly (100 deg C per minute) so that the glass does not devitrify. High fabrication temperature and rapid cooling can lead to flaws which result in interfacial failure. The frit composition is selected to match expansion and provide wetting and adhesion to both the metal and ceramic. The bond can be heated much higher than an SMPP bond - on the order of 800 deg. C. Nonmetallic fusion has been used to bond high purity alumina and sapphire successfully.

5. Gas-Metal Eutectic Method (Direct Bonding)

The gas-metal eutectic method, also termed direct bonding, has been growing in application. It can be used to bond oxide ceramics such as aluminum oxide, silica, beryllia, zirconia and spinels as well as non-oxide ceramics such as aluminum nitride, silicon nitride and silicon carbide to such metals as copper, nickel, iron, molybdenum and their alloys (24, 25, 26). The metal can quickly be bonded directly to the ceramic without intermediary material, with little or no pressure and without melting the metal.

The method takes advantage of the fact that several metals show a low melting eutectic with their oxide (sometimes sulfide or silicide). For example, at low partial pressures of oxygen copper has a eutectic with copper oxide below the melting point of copper. Copper can be preoxidized and bonded in a vacuum or inert atmosphere or alternatively oxidized while bonding in a low partial pressure of oxygen. The metal/metal oxide melts while the bulk metal remains solid. The process requires careful atmosphere and temperature control to avoid melting the bulk metal shape. The eutectic melt wets and bonds to the ceramic. It has been demonstrated recently that in the case of copper bonded to alumina a reaction takes place to form a further interfacial bonding phase, copper aluminate. This is a low melting eutectic in the tertiary system and shows that a further bonding reaction may be important (26). The minor element composition of the ceramic has also been shown to have major effects on bond strength (26). This method has also been successfully applied to bonding non-oxide ceramics such as aluminum nitride by first pre- oxidizing to form alumina on the ceramic surface.

6. Pressed Diffusion Method (Direct Bonding)

Joining by solid phase diffusion and reaction may be employed to bond metals and ceramics. High strength bonds require careful surface preparation, elevated temperature, long times and high pressures. Such bonds are termed pressure-diffusion, ram or crunch joins. This approach can be particularly suitable for low yield stress metals such as copper. Copper can be pressed to conform to the ceramic or the ceramic forced into a cavity in the metal. This can give good contact between the metal and ceramic, with a residual stress at their interface. The parts must be as clean as possible before contact. The parts are held with a high external force applied normal to the interface at as high a temperature as possible, but above half the melting temperature of the metal. The gas can have a profound effect on joining both in terms of trace impurities and different effectiveness with inert gas chosen. A variety of metals and ceramics can be coupled by this method (27, 28).

7. Vapor Phase Metallizing

A thin metallic film can be applied by various deposition methods from the vapor phase. Such methods include: evaporation, sputtering, flame spraying, reactive sputtering, chemical vapor deposition (CVD), plasma spraying and ion implantation. Frequently an active metal is deposited such as titanium, zirconium and chromium to promote bonding, followed by noble or base metals as for example gold, platinum, palladium, nickel and copper. Final layers are usually oxidation resistant metals. Platinum shows similar thermal expansion, good ductility and excellent adhesion to alumina ceramics and is frequently used directly. Often a very tenacious bond is developed with thin film vapor phase applications for metals which bond poorly as bulk material. There may be rather different bonding when active atoms or ions are deposited as compared to liquid or solid phase bonding. Contact can occur atom-by-atom providing superior physical bonding, even in systems which do not wet. The mechanical behavior of thin films also differs from bulk material and may also account for the tenacity of many bonds.

Usually metallization applied by vapor phase methods are quite thin, although layers several micrometers can be developed. Plasma spraying and CVD techniques are most used to build up thicker structures. After deposition, layers can be built up by conventional solder and braze methods or alternatively electroforming can be used for joining to metal parts after vapor phase deposition has been completed. High temperature brazing is typically done in hydrogen, under vacuum or with a cracked ammonia atmosphere. Ceramics and glasses not tolerant of brazing temperatures can be soldered at lower temperatures. Strong, hermetic seals can be made on glasses, single phase, polyphase and single crystal ceramics.

A wide variety of methods are available for applying ceramic films to metals including thermal spraying, plasma deposition, radio frequency [RF] and reactive sputtering, chemical vapor deposition [CVD] and ion implantation. These coatings can be used in in themselves. Such coatings are also used occasionally as the base upon which to bond a bulk ceramic. In such applications a strong bond is required between the base metal and the primary ceramic coating. This can be best achieved by vapor techniques with higher energy inputs such as plasma spraying and reactive and/or RF sputtering. These techniques provide a rather intimate and potentially strong bond to the metal which may then function as a joining phase for successive layers or a bulk joint. Thermal mismatches must be taken into account for successful application as even when the application is made on a small area the high energy input can affect bond quality.

8. Liquid Phase Metallizing

A salt solution of a precious metal (platinum, gold, silver, palladium) is applied to the ceramic and then thermally reduced. A thin conductive coating results which can be soldered, brazed or electroplated directly. Reactive metals can be included in the composition in order to improve adherence. This method is particularly handy for putting electrodes on ceramics for electrical characterization and for very low volume applications.

9. Electroforming

After a ceramic part has been metallized by vapor or liquid phase techniques to provide a thin conductive layer, electroforming can be used in order to avoid high temperature brazing. The metallized ceramic and metal (often copper) are pressed together and the assembly placed in an electroplating bath for some days. Electrodeposition in the gap bonds the metal and metallized ceramic. Large, complex parts can be joined this way avoiding thermal expansion mismatch stresses. The method can be coupled with pressure diffusion joining techniques, but this no longer avoids elevated temperature and thermally induced stresses on cooling.

10. Graded Powder Process

The graded powder process is frequently closely related to either the metal powder-glass frit process or the SMPP process. A powder compact is made in which the composition varies in layers from ceramic-rich to metal-rich. This mechanically produced variation in the preform permits a more ceramic-like bond material at the ceramic interface and greater metal content for joining to the metal. The composition, properties and structure is graded between the ceramic and metal. The powder is fired at a temperature similar to the SMPP or metal powder-glass frit upon which it is based. Because the bond is only slightly stronger than other methods and the preform increases costs only limited application has taken place.

Acknowledgments

Support by the Center for Ceramic Research, Rutgers University, its industrial sponsors and the New Jersey Commission on Science and Technology is gratefully appreciated.

References

1. J.T. Klomp, Interfacial Reactions Between Metals and Oxides During Sealing, Bull. Am. Ceram. Soc., Vol 59, 1980, p 794-800.

2. M.E. Twentyman, High Temperature Metallizing, J. Mater. Sci., Vol 10, 1975, p 765-766.

3. W.H. Kohl, Handbook of Materials and Techniques for Vacuum Devices, Reinhold, 1967.

4. J.R. Tinklepaugh and W.B. Crandall, Ed., Cermets, Reinhold, 1960.

5. J.A. Pask and R.M. Fulrath, Fundamentals of Glass-to-Metal Bonding: VIII, Nature of Wetting and Adherence, J. Am Ceram. Soc., Vol 45, 1962, p 592-596.

6. F. Bashforth and S.C. Adams, An Attempt to Test Theories of Capillarity, CAmbridge University Press, 1883.

7. A.G. Pincus, Mechanism of Ceramic-to-Metal Adherence, Adherence of Molybdenum to Alumina Ceramics, Ceram. Age, Vol 63, 1954, p 16-20, 30-32.

8. J.D. Partridge, Glass-to-Metal Seals, Society of Glass Technology, 1949.

9. D. Tabor, Interaction Between Surfaces: Adhesion and Friction, Surface Physics of Materials, Vol 11, Academic Press, 1975, p 475-529.

10. M. Borom and J. Pask, Role of Adherence Oxides in the Development of Chemical Bonding at Glass-Metal Interfaces, J. Am. Ceram. Soc., Vol 49, 1966, p 1-6.

11. H.E. Pattee, R.E. Evans, and R.E. Monroe, Joining Ceramics and Graphite to Other Materials, WASH SP-5052, National Aeronautics and Space Administration, 1968.

12. G.R. Van Houten, A Survey of Ceramic-to-Metal Bonding, Ceram. Bull., Vol 38, 1959, p 301-307.

13. G.P. Chu, Some Aspects on Glass-Ceramics to Metal Sealing, in Proceedings of the Tenth International Congress on Glass, Ceramic Society of Japan, 1967, p 94-102.

14. C.I. Helgesson, Ceramic-to-Metal Bonding, Boston Technical Publishers, 1968.

15. S.S. Cole and G. Sommer, Glass Migration Mechanism of Ceramic to Metal Seal Adherence, J. Am. Ceram. Soc., Vol 44, 1961, p 265-271.

16. R.E. Loehman, Interfacial Reactions in Ceramic-Metal Systems, Bull. Am. Ceram. Soc., Vol 68, 1989, p 891-896.

17. R.G. Loasby, N. Davey, and H. Barlow, Enhanced Property Thick Film Pastes, Solid State Technol., Vol 15 (No. 5), 1972, p 46-72.

18. M. Monneraye, Les Encres Sengraphiables en Microelectronique Hybride-les Materiaux et Leur Comportement, Acta Electron., Vol 21, 1978, p 263-281.

19. C.S. Pearsall, New Brazing Methods for Joining Non-Metallic Material to Metals, Mater. Methods, Vol 30, 1949, p 61-62.

20. J.E. McDonald and A. Eberhart, Adhesion in Metal Oxide Systems, Trans. Metall. Soc. AIME, Vol 233, 1964, p 512-517.

21. J.J. Schmidt and J.L. Carter, Using Nitrogen Based Atmosphere for Glass-to-Metal Sealing, Met. Prog., Vol 128, 1985, p 29-34.

22. S. Tamhankar, N. Bandyopadhyay, and M. Kirschner, Glass-to-Metal Sealing by a Single-Atmosphere Single-Firing Process, Int. J. Hybrid Microelectronics, Vol 11, 1988, p 31-36.

23. N. Bandyopadhyay, S. Tamhankar, M. Kirschner, Glass-to-Metal Sealing Process, U.S. Patent 4,888,037 (1989).

24. J.F. Burgess, C.A. Neugebauer, and G. Flanagan, The Direct Bonding of Metals to Ceramics by the Gas-Metal Eutectic Method. J. Electrochem. Soc. Solid-State Sci. Technol., May 1975.

25. J.T. Klomp and P.J. Vrugt, Interfaces Between Metals and Ceramics in Surfaces and Interfaces, Mater. Sci. Res., Vol 14, 1980, p 97-105.

26. J.E. Holowczak, V.A. Greenhut, and D.J. Shanefield, Effect of Alumina Composition on Interfacial Chemistry and Strength of Direct Bonded Copper-Alumina, Cer. Eng. Sci. Proc., Vol 10, 1989, p 1283-1294.

27. W. Dawihl and E. Klinger, Mechanical and Thermal Properties of Welded Joints Between Alumina and Metals, Ber. Dtsch. Keram. Ges., Vol 46, 1969, p 12-18.

28. H.J. de Bruin, A.F. Moodie, and C.E. Warble, Reaction

Welding of Ceramics Using Trensition Metal Intermediates, J. Aust. Ceram. Soc., Vol 7 (No. 2), 1971, p 57-58.

Aluminum/Alumina Joining

R. A. Haber and V. A. Greenhut

Department of Ceramics
Rutgers University
Piscataway, NJ 08855-0909

Abstract

Aluminum tensile adhesion to alumina composites containing varying volume fractions of borosilicate glass was examined. Alumina composites were found to bond to aluminum at a reaction temperature of 800°C with tensile bond strengths of 62 MPa observed for composites containing 15% glass.

Introduction

Metal-ceramic particle and fiber composites have been widely recognized for wear resistance and high strength potentials. The types of composites vary widely depending upon application, but it is generally agreed that alumina fibers or particles are ideal structural reinforcing agents for aluminum and its alloys. However, there are a number of major difficulties encountered in fabricating aluminum-alumina composites by melt metallurgy. These include the absence of wetting and bonding between molten aluminum and alumina at temperatures used in conventional foundry practices, and the segregation of dispersed particles or discrete fibers in the melt and castings due to differences in density between particles and metal.[1-6]

Numerous approaches have been employed to improve the wetting between molten aluminum and alumina. Fleming and Mehrabian, and Quigley et al. found that by mechanically stirring molten alumina fibers into molten aluminum, alumina fibers could be wet by the metal.[2,6] However, this technique required highly specialized stirring equipment and careful temperature control.

Champion et al. modified the aluminum alloy by additions of small amounts of active metals such as lithium, calcium or magnesium to promote wetting.[7] Munitz et al.[8] and Quigley et al.[2] found that magnesium alloying additions enhanced the

Metal - Ceramic Joining
Edited by P. Kumar and V. A. Greenhut
The Minerals, Metals & Materials Society, 1991

development of a $Mg_2Al_2O_4$ spinel at the alumina interface which aided in the development of a strong interface between alumina and aluminum.

Anderson showed that by using an in-situ oxidation/infiltration process (Lanxide Process) both continuous alumina fibers and particulate alumina could be coated by aluminum and made into composites. This process required careful control of temperature, atmosphere, and alloying additives.[9]

Only two investigations of the adhesion strength between aluminum and alumina have been reported. Nicholas et al. determined tensile bond strengths ranging from 11-67 MPa. These strengths were reported for varied alloying, i.e. lithium and magnesium, additions made to aluminum on 99% alumina.[10,11] In all cases reaction temperatures were in excess of 1100°C; 1120°C for low strengths and 1390°C for high strengths. Wolf, Levitt and Brown reported shear bond strengths of 43 MPa) for pure aluminum and 89 MPa for aluminum with selenium alloying additions to high purity alumina.[12]

This investigation will show that interfacial reactions resulting in strong adhesion between aluminum and alumina can be obtained at low temperatures, 800°C, without altering the metal, but rather by changing the character of the ceramic by controlled additions of a borosilicate based glass to the alumina.

2. Experimental

To analyze the interfacial bond strength between Al and Al_2O_3/glass composites, a tensile bond test configuration was adopted. This required specially fabricated ceramic substrates in addition to unique metal/ceramic test specimens.

The aluminum utilized for this evaluation was high purity, 99.99+% Al. It was obtained in the form of a small laboratory ingot and sawed into small pieces for melting.

The aluminum metal was cut into 2.5 cm inch cubes. These cubes were then cleaned by washing the cubes in a NaOH, Na_3PO_4 and water solution for 10 minutes, followed by a rinse in water, a subsequent soak in a HF, HNO_3 and water solution for 10 minutes, and finally to rinsed in water and ethanol. The cleaned cubes were then placed into a closed bottom graphite crucible and heated to 650°C, see the upper crucible in Figure 1. The metal was then allowed to cool until required for testing.

The alumina/glass composites were fabricated by prereacting A-16SG(Alcoa, Pittsburgh, PA) alumina and borosilicate glass frit (Pemco P54 Frit, Baltimore, MD) in varied proportions. For this investigation the glass varied from 5 to 40 Vol. % additions.

The ceramic composites were prepared by wet mixing the proportioned powders in water and filtering the suspensions into a cake. The cake was then broken into 1-2 mm granules dried and fired to temperatures ranging from 1100-1600°C depending on the glass content, see Table 1. The prefired granules were ground by ball milling to 1-5 μm size. The resulting powder was slip cast into 7 x 7 x 0.5 cm tiles, dried and fired a second time to temperatures ranging from 1150-1625°C. The fired tiles were polished to a 1 μm finish and sectioned into 2 x 2

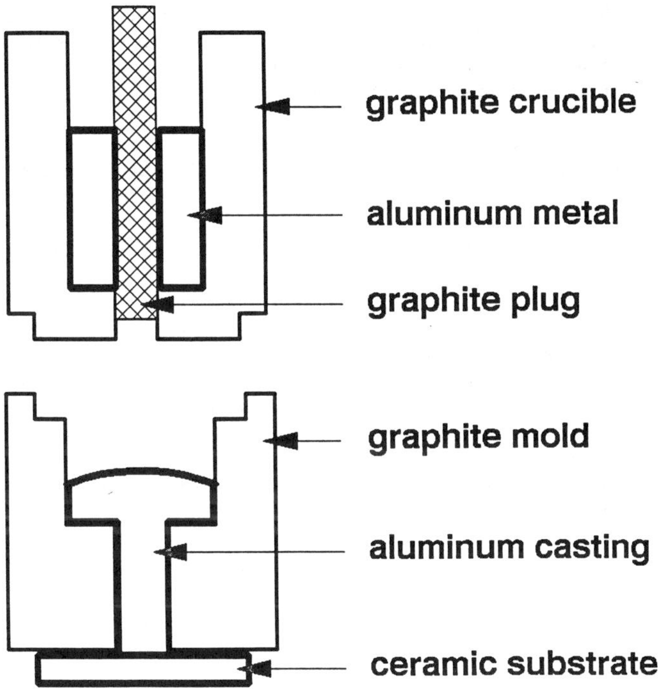

Figure 1. Graphite crucible and mold for casting tensile bond strength specimen.

cm squares for testing. Figures (2a and b) show the typical microstructure for the fired ceramics. In both cases the microstructures are dense with the exception of some polishing pull-outs.

A casting system was developed whereby molten metal was contained in a graphite crucible and cast onto a substrate through a bottom spigot in the containment crucible. The casting was contained within a graphite mold which predefined, 0.6 cm diameter, the metal-ceramic contact area. This configuration prevented the formation of an oxide skin on the metal since any oxide of the metal present will tend to float.

The graphite crucibles and molds were machined to specified shapes shapes. The graphite crucible containing the previously cast aluminum blank, was fit onto the mold. The sample assembly was placed into electrically heated vertical tube furnace. The furnace was configured to allow gas to flow and permit the graphite plug to be removed from the metal casting crucible at the $800^{o}C$.

The casting operation entailed flushing the furnace with argon and then heating the furnace to $800^{o}C$ at a rate of $300^{o}C$/hr. Once the assembly reached temperature, the graphite plug was removed allowing the molten metal to run through the hole in the bottom of the crucible and fill the graphite mold. The metal remained in contact with the ceramic for two time periods, 1 and 3 hours. After this, the furnace was cooled at a rate of $3^{o}C$/minute to $30^{o}C$. The graphite mold was taken apart and the sample of aluminum cast directly upon the Al_2O_3/glass composite removed.

Interfacial adhesion strengths were measured in a uniaxial tensile configuration. Sample configuration consisted of the polished flat ceramic substrate and a metal casting in the shape of a mushroom, see Figure 3. To evaluate interfacial adhesion the metal mushroom shaped casting was pulled from the substrate using a tensile testing machine with universal joints, see Figure 4. The crosshead speed was 0.05 cm/min.

Discussion of Results

The results of the tensile bond strength tests are shown in Figure 5. All tests were performed at samples cast at $800^{o}C$. The data shows that longer reaction times resulted in higher tensile adhesion and that maximum strength corresponded to 15 to 20 Vol. % glass additions to the composite.

Figures 6a and 6b show typical fracture surfaces of 70 and 95 wt. % alumina composites, respectively. These compositions provide microstructures representative of the low and high volume percent glass containing composites used. It is apparent that there are numerous changes that occurred that led to bonding. Figure 6a does not show any evidence of metal. While the ceramic surface appears roughened, most likely from a reaction of the metal with the small amount of glass present, little bonding took place. Figure 6b shows strong evidence of bonded metal noted by the ductile stretching regions corresponding to the aluminum. Also evident is the highly corroded appearance of the ceramic composite in near proximity to the metal regions. X-ray fluorescent analysis showed these regions to be depleted in boron from the glass.

TABLE 1

Volume Percent Alumina in Composite	Prereaction Temperature °C	Second Fire Temp. °C
99	1600	1625
95	1500	1550
85	1425	1475
75	1300	1350
70	1200	1250
60	1100	1150

Prereaction and second fire temperatures for the ceramic composites. Firing temperatures are based on a 3°/min heat up ramp, a 1 hour soak at temperature and a 3°/min cool down ramp.

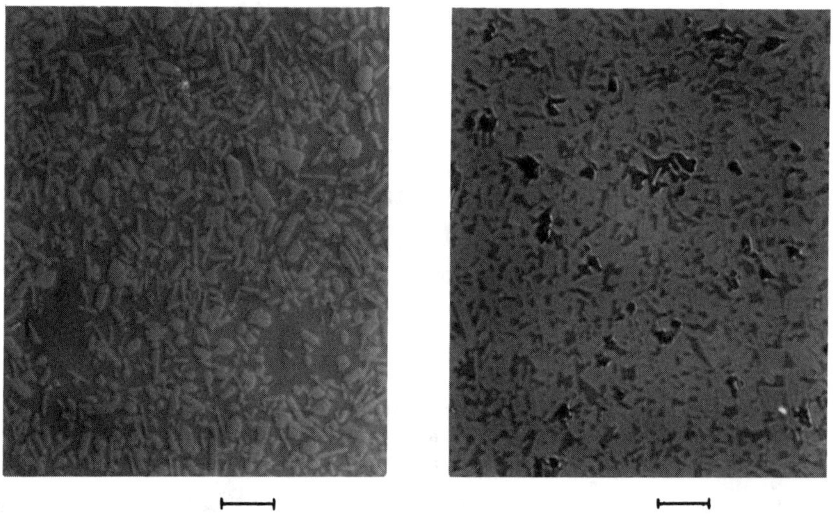

(a) (b)

Figures 2a and b illustrate typical dense microstructures of the ceramic substrates. Figure 2a shows a 80% alumina, 20% glass composite. Figure 2b shows a 90% alumina, 10% glass sample. In both samples the lighter regions are the particulate, the gray region glass and the dark regions porosity originating as pull-outs.

Figure 3. Cross-section of metal casting mold with tensile specimen.

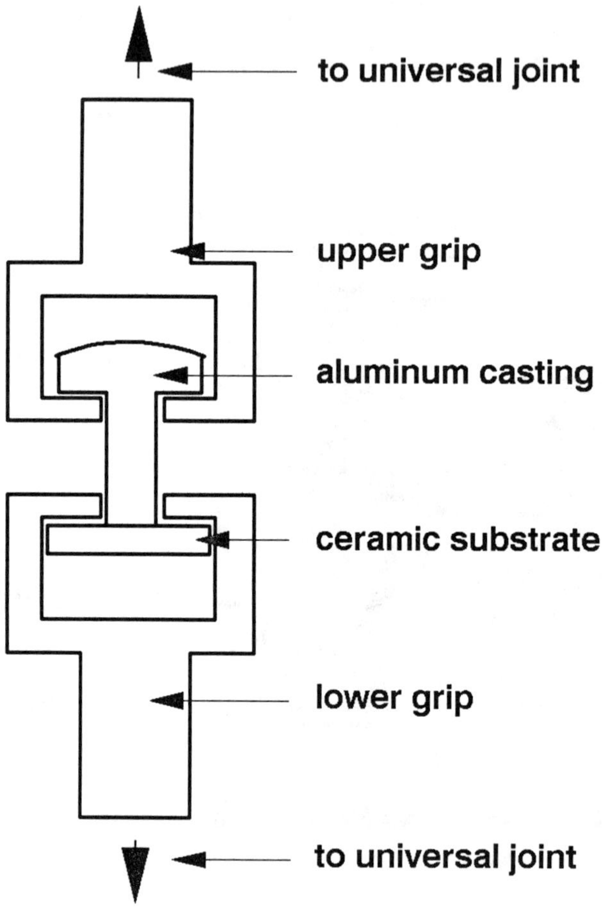

Figure 4. Tensile testing fixture coupled to a Instron testing machine.

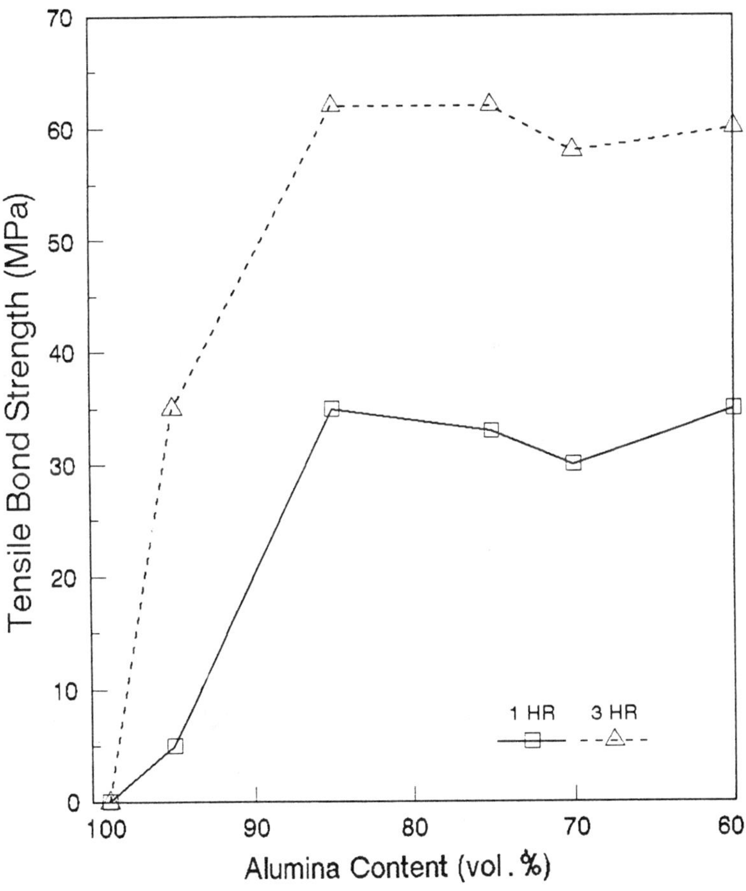

Figure 5. Tensile bond strength for cast aluminum on alumina/glass composite as a function of metal/ceramic contact time.

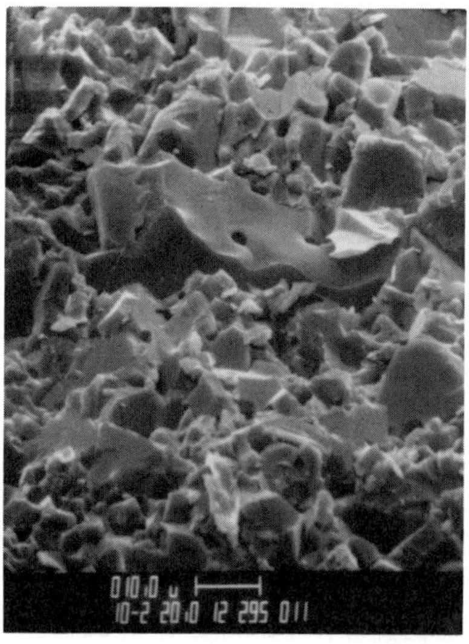

(a)

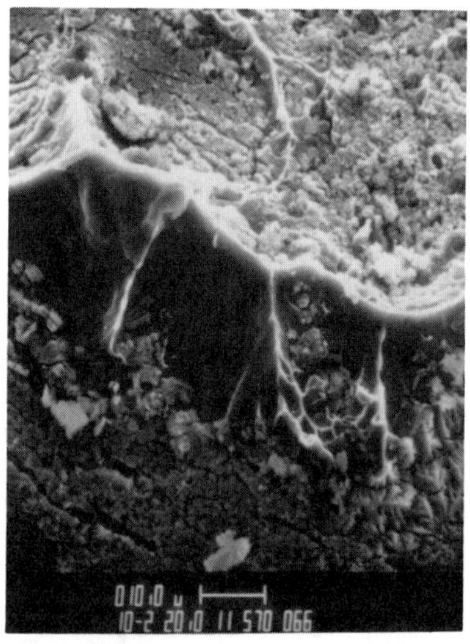

(b)

Figures 6a and 6b show the surface of the ceramic composite after tensile testing for a 95% alumina, 5% glass and 70 % alumina, 30% glass composites, respectively.

It appears that the molten metal wets and reacts with particulate areas. It is in these areas where the metal obtained a sufficient mechanical interlocking or key-holing that ductile stretching or ductile dimpling was observed. A striking feature of these micrographs is the nature of the surface where aluminum was and was not in contact with the ceramic. The areas where metal was in contact are depleted of fine grained alumina particles and their surrounding glassy phase. Whereas the uncontacted areas exhibit distinct fine grain size particles and an accompanying glassy phase. This is well observed in Figure 7. As was expected from prior studies by Wolf et al., poor wetting occurred when molten aluminum was placed in contact with alumina containing small amounts of glass. This resulted in a lower tensile adhesion. Whereas, alumina composites containing higher levels of glass exhibited wetting, aluminum to glass, resulting in interfacial bonding. The well distributed glassy phase within the ceramic provided a large number of bonding regions. As the time the metal was left in contact with the composite increased, so did the reaction with the glass, which resulted in higher tensile strengths.

The differences in tensile bond strengths can be accounted for by a combination of the chemical and mechanical bonds. In high alumina composites there is not a sufficient amount of glass present to develop strong chemical interactions. As the volume fraction of glass increases the degree of chemical bonding and accompanying mechanical bonding increases. At 15-20 Vol. % glass present an optimum bond develops which reflects both strong chemical and mechanical interactions. However, as the glass in the ceramic composite increases beyond 20 Vol.%, the amount of available alumina for mechanical anchoring decreases. Thus, the resulting bond reflects high degrees of chemical interaction and very little mechanical anchoring and the metal-ceramic bond strength decreases.

Summary

The nature of the bond that develops between aluminum and alumina/glass composites utilizes two basic ceramic-metal joining concepts; mechanical or keyhole joining and chemical or interfacial joining. When aluminum is bonded directly to alumina, no significant reaction or bond is observed. When glass is incorporated in to an alumina microstructure as a highly continuous second phase, metal-ceramic composite bonds in excess of 60 MPa were achieved. The glass was formed to form a transitional bond between the ceramic and the aluminum.

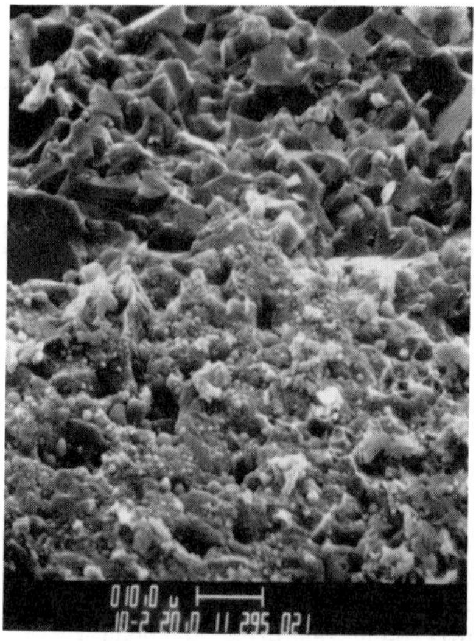

Figure 7. A comparison of the unreacted to reacted area left in contact with aluminum, (lower portion) for a ceramic composite containing 5 Vol.% glass.

References

1. Wolf, S.M. and Levitt, A. P., "Whisker-Matrix Bonding and Interaction," AMMRC-MS, 68-11, Watertown, Mass. (1968).

2. Quigley, B. F., Abbaschian, G. J., Wunderlin, R. and Mehrabian, R., "A Method of Fabrication of Aluminum-Alumina Composites," Met. Trans., 13A, 93-100 (1982).

3. Brennan, J. J. and Pask, J. A., "Effect of Nature of Surfaces of Wetting of Sapphire by Liquid Aluminum," J. Am. Cer. Soc., 51 (10), 569-73 (1968).

4. Carnahan, R. D., Johnston, T. L. and Li, C. H., "Some Observations on the Wetting of Al_2O_3 by Aluminum," J. Am. Cer. Soc., 41 (9) 343-47 (1958).

5. Mehan, R. C., "Fabrication and Mechanical Properties of Sapphire-Whisker-Aluminum-Ceramic Particle Composites," J. Mat. Sci., 16, 983-993 (1981).

6. Goddard, D. M., "Interface Reactions During Preparation of Aluinum-Matrix Composites by the Sodium Process," J. Mat. Sci., 13, 1841-48 (1978).

7. Champion, A. R., Krueger, W. H., Hartman, H. S. and Dhingra, A. K., "Fiber FP Reinforced Metal Matrix Composites," Proc. Second International Conference on Composite Materials, p. 883, Met. Soc. of AIME (1978).

8. Munitz, A., Metzger, M. and Mehrabian, R., "The Interface Phase in Al-Mg/Al_2O_3 Composites," Mef. Trans., 10A, 1491-97 (1979).

9. Anderson, Lanxide Corp., Newark, DE, "Private Communication".

10. Nicholas, M., Forgan, R. R. D. and Poole, D. M., "The Adhesion of Metal/Alumina Interfaces," J. Mat. Sci., 3, 9-14 (1968).

11. Nicholas, M., "The Strength of Metal/Alumina Interfaces," J. Mat. Sci., 3, 571-76 (1968).

12. Wolf, S. M., Levitt, A. P. and Brown, J., "Wetting and Bonding Between Aluminum Alloys and Sapphire," AMRA-TA, 66-79, Watertown, Mass. (1966).

13. Haber, R.A., "The Effect of Controlled Glass Additions on the Strength of High Alumina Ceramics", Ph.D. Thesis, Rutgers University, New Brunswick, NJ, 1984.

RESIDUAL STRESS ANALYSIS AND MICROSTRUCTURAL OBSERVATIONS OF CERAMIC-TO-METAL BRAZED JOINTS

A. Levy, A. Tobin, and G. Busch

Grumman Corporate Research Center
Bethpage, New York 11714-3580

Abstract

Brazing is commonly used to join ceramics to metals. As the joint is cooled from the brazing temperature to ambient temperature, the interface microstructure changes and residual stresses develop. In this study we characterize these developments and explore their effects on the integrity of the brazed joint. A wide range of materials with differing mechanical, physical, and thermal properties were utilized. Ceramic materials included Al_2O_3, SiC, while metallic materials included Ti, W, Ti_3Al, and Ta. A silver-based braze alloy that contained titanium to provide good wetting to both metallic and ceramic surfaces was utilized to create the interface. Brazing was performed in a vacuum furnace for various times and temperatures. The effects of brazing on the interface alloy microstructure and redistribution of alloying elements in the interfacial region were experimentally investigated using both optical and SEM/EDS techniques. In addition, residual stresses develop as the material is cooled from the brazing temperature to ambient temperature, due primarily to the mismatch in coefficients of thermal expansion of the joined materials. These residual stresses affect the properties and strength of the brazed system and in some cases lead to catastrophic failure at or near the joints during the brazing process. Utilizing both an approximate elastic closed-form solution and more accurate elastoplastic finite element methods that take into account the temperature dependent stress-strain response of the braze material, residual stresses were determined for various ceramic-metal combinations. These results helped to explain the causes of joint failure. The effects of the braze material work-hardening rate and joint thickness on the level of residual stress were investigated. Analysis of the practical case in which ceramic tiles are brazed onto a metal substrate, led to different results than those found in the laboratory brazed coupon. Qualitative agreement exists between analytical predictions and experimental findings regarding brazing failure modes.

Metal - Ceramic Joining
Edited by P. Kumar and V. A. Greenhut
The Minerals, Metals & Materials Society, 1991

1. Introduction

Because of their low density, excellent oxidation behavior, erosion and thermal shock resistance, and retention of mechanical properties at elevated temperatures, ceramic materials are attractive for application to high heat flux areas of advanced energy systems such as fusion reactors and hypersonic vehicles. Attachment of these materials to the primary structure of a vehicle is a critical problem, since ceramics are very sensitive to localized thermal and mechanical stresses associated with any attachment scheme. Brazing is a potentially attractive method of fabricating a ceramic-to-metal joint since localized stresses associated with mechanical attachments are avoided and excellent thermal contact between the ceramic and metal can be achieved.

The objective of this research is to investigate critical factors that control the formation of a ceramic-metal joint formed via brazing. The areas of investigation include the microstructural compatibility between the phases being joined (including the filler metal, referred to as the braze) and the residual stresses developed during fabrication. The effects of brazing on the interface alloy microstructure, and redistribution of alloying elements in the interfacial region were experimentally investigated, using both optical and SEM/EDS techniques. When brazing ceramic to metal, residual stresses are developed as the material is cooled from the brazing temperature to ambient temperature. This is primarily due to the mismatch of coefficients of thermal expansion between the individual materials. The residual stresses may affect the properties and strength of the joint during mechanical or thermal loading and, in some cases, lead to catastrophic failure during the brazing process. Recent work by Charreyron, et al., (1) and Patten, et al., (2) using finite element analysis (FEA) to compute stress intensity factors, has shown that stresses developed in a brazed ceramic-metal joint can lead to nucleation and propagation of cracks near the free edges of a brazed bond. Plane stress is assumed, which applies for thin adherents, and the Dugdale model is used to approximate the plastic behavior of the braze. Cao, et al., (3) investigated the residual stress when two ceramic materials are bonded with a thin ductile metallic layer. Evans, et al., (4) and Dalgleish, et al., (5) have investigated the mechanical strength of ceramic-metal bonded systems, but without taking into account the residual stresses developed during the cooling process. However, both point out that the residual stresses may significantly influence the strength of a bonded system.

In the present paper, analytic studies of the stresses in the braze joint were carried out. First, we obtained an approximate closed-form solution of the elastic stresses as a function of geometry, thermal expansion coefficients, elastic moduli, and temperature change. This approximate solution allowed qualitative insight into the joint residual stresses. However, to obtain a more accurate prediction of the stress distribution in the brazed joint, FEA methods were employed. In our analysis, the elastoplastic properties of the braze alloy, including work-hardening effects and temperature dependent braze stress-strain responses, were accounted for.

Various ceramic-metal combinations were considered. Incusil ABA, a commercial Ag-Cu-In-Ti alloy (WESGO, Belmont, CA) was utilized as the joining material. The effect of Incusil work-hardening rate and thickness was investigated. Also, we distinguished between the laboratory brazed ceramic-metal coupon and the practical case in which ceramic tiles are brazed onto a metallic substrate.

Fracture criterion should be based on stress intensity factors or energy release rates. Nevertheless, investigating stress levels along the potential failure interfaces are extremely helpful, in that these residual stresses are the driving forces of the joint failure. Qualitative agreement exists between analytical predictions and experimental findings regarding failure modes.

2. Approach

2.1 Experimental

All ceramic, metallic, and braze alloy materials were obtained from commercial sources. Table I lists all the materials used in this study. All ceramic materials were fully densified via either sintering or hot pressing, and they exhibited little or no porosity. Ceramic materials were cut into 2.5x2.5-cm squares and degreased in trichloroethylene and acetone prior to bonding. Metallic substrates were obtained from commercial sources in the form of sheet or plate and cut into 2.5x2.5-cm squares, degreased in acetone and acid etched prior to bonding. Incusil ABA, a Ag-Cu-In-Ti braze alloy, in the form of 25x25x0.074-mm foil, was utilized as the joining material. This silver-copper-indium braze alloy has a brazing range of 750-800°C and contains a small concentration of titanium to assist wetting of the ceramic. Excellent wetting of the braze to all ceramic materials studied was observed. To assure good contact between parts being joined, a 1-lb steel weight was place on top of the assembly during the brazing operation. Brazing was conducted in a vacuum oven (10^{-5} torr) heated to 800°C for periods of 5-15 minutes followed by a furnace cool. Samples were inspected after brazing for visible presence of cracks in the ceramic phase. Brazed composites were sectioned across their interfacial zone and polished. Both optical and SEM/EDS techniques were used to study the microstructure and elemental distribution of the interfacial region. Backscattered electrons were used to distinguish amongst localized regions of differing average atomic number within the braze and EDS mapping techniques were used to map elemental distributions.

Table I. Ambient Temperature Material Properties

Material	α ($\times10^{-6}$/°C)	E, GPa	ν	E_p, GPa	σ_{ys}, MPa
Ceramic					
SiC	4.5	400	.24		
Al_2O_3	8.1	372	.27		
Braze					
Incusil	19.7	76	.33	4.75	235
Metal					
Ti	8.5	114	.33		
W	4.6	380	.3		
Ta	6.6	176	.3		
Ti_3Al	8.6	124	.31		

E is Young's modulus, E_p is the slope in the plastic region of a bilinear representation of the stress-strain curve, ν is Poisson's ratio, and σ_{ys} is the yield stress.

2.2 Analytical

Three basic models, depicted in Fig.1, were analyzed. In all cases a thin slab has been cut away from the transverse free edge (perpendicular to the y direction) so that in the transverse direction we have a plane strain condition. In the width direction, x, the model need only go from the free edge to a distance at which the edge effect is not felt and the material behaves in plane strain (axes are shown in Fig. 3).

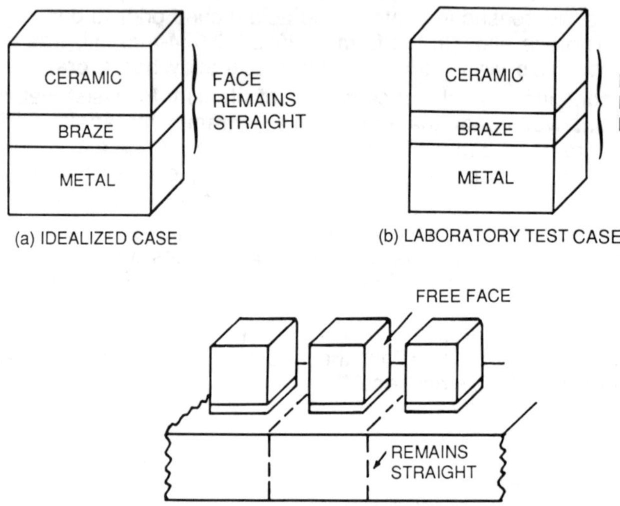

(c) CERAMIC TILES BRAZED ONTO METAL SUBSTRATE

Fig. 1 Brazed Ceramic-to-Metal Models

2.3 Simplified Elastic Solution

Figure 1a corresponds to an idealized case in which the free faces in the width direction are constrained to remain plane. The in-plane forces for each layer due to a temperature change, ΔT, are expressed as (see, for example, 6)

$$F_i = W\, t_i\, E_i\, [\, \Sigma_j\, E_j\, t_j\, (\alpha_j - \alpha_i) / \Sigma_j\, E_j\, t_j]\, \Delta T\,, \quad \Sigma F_i = 0,\ i,j = 1,n \qquad (1)$$

where W is the width (x direction), t_i the layer thickness, E_i the layer Young's modulus, and α_i the layer coefficient of thermal expansion, and n is the number of layers.

The average shear stress at an interface is approximated as

$$\tau = (\Delta F/2)/(LW) \qquad (2)$$

where $(\Delta F/2)$ is the average change in force across the interface, computed from Eq. (1), and L is the distance over which the shear stress builds up (i.e., edge effect) and is unknown. By approximating L we can obtain an approximate solution, from Eq.(1), for the average shear stress at each interface.

2.4 Finite Element Analysis

The material properties for the ceramic, metal, and braze, are a function of temperature. During cooling from the braze temperature to ambient temperature, Young's moduli increase and the coefficients of thermal expansion decrease. From a simple elastic solution for thermal residual stresses in a layered system - Eq. (1) and Eq. (2) - the shear stresses developed at the joint interfaces are approximately dependent on the differences between the individual materials': (1) coefficient of thermal expansion and (2) product of Young's modulus and coefficient of thermal expansion. These differences are much less dependent on temperature than the individual material properties (Young's moduli and coefficients of thermal expansion); therefore, little is lost by considering only ambient temperature properties. Ambient temperature properties for the materials under consideration are given in Table I. However, the stress-strain response of the Incusil braze drastically changes with temperature, as shown in Fig. 2, and was accounted for in our analysis.

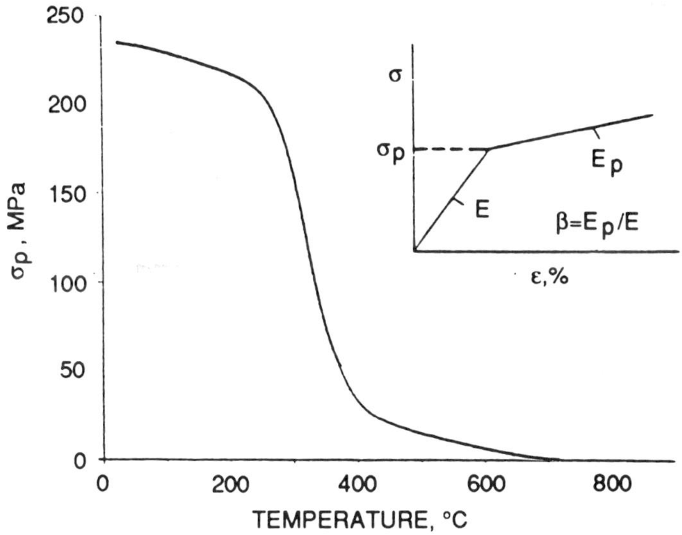

Fig. 2 Temperature Dependent Proportional Limit for Incusil Braze

Two cases, depicted in Figs. 1b, and 1c are analyzed. Figure 1b corresponds to a brazed ceramic-to-metal laboratory test coupon and Fig. 1c corresponds to the case in which ceramic tiles are brazed onto a metal substrate. In both cases a thin slab has been cut away in the transverse direction (parallel to the xz plane). This corresponds to a plane strain condition in the y direction. In the x direction the model need only go from the free edge to a distance at which the edge effect is not felt. At this distance plane faces remain plane (this is designated as x=0). The ceramic and braze faces are stress free and the metal faces are stress free in the laboratory coupon case (Fig. 1b) and constrained to remain planar in the case of ceramic tiles brazed onto a metal substrate (Fig. 1c) (this corresponds to x=1000 μm in all subsequent result figures). The thicknesses of the ceramic, braze, and metal layers are 250 μm, 75 μm, and 500 μm, respectively.

We will refer to the finite element grid, shown in Fig. 3, as our "standard" grid. This grid contains 288 isoparametric elements with midside nodes. An incremental numerical procedure is utilized (7,8) based on the von Mises yield criterion, the Prandtl-Reuss flow rule, and the Prager-Ziegler kinematic work-hardening theory, which accounts for the Bauschinger effect. In addition, modifications to account for a temperature dependent yield stress and corresponding constitutive equations (9) are included.

Under certain conditions a stress singularity exists at the ceramic/braze or metal/braze interface, near the free edge. The accuracy of FEA is well established, except near a stress singularity or discontinuity. This is because the assumption of a symmetric shear stress does not hold. The stress tensor can become unsymmetrical if the first partial derivative of stress is unbounded (10). Whitcomb, et al., (11) investigated the accuracy close to a singularity. A bimetallic plate subjected to uniform tension was used to illustrate a singularity near the free face of intersecting dissimilar materials. They concluded that valid results may be obtained by FEA everywhere except in the two elements closest to the singularity, and that this region can be made arbitrarily small by mesh refinement. In addition, the grading of element size, rather than using equal size elements, is well suited for regions surrounding a singularity . In our present work we explore the acceptability of our "standard" grid to produce accurate results.

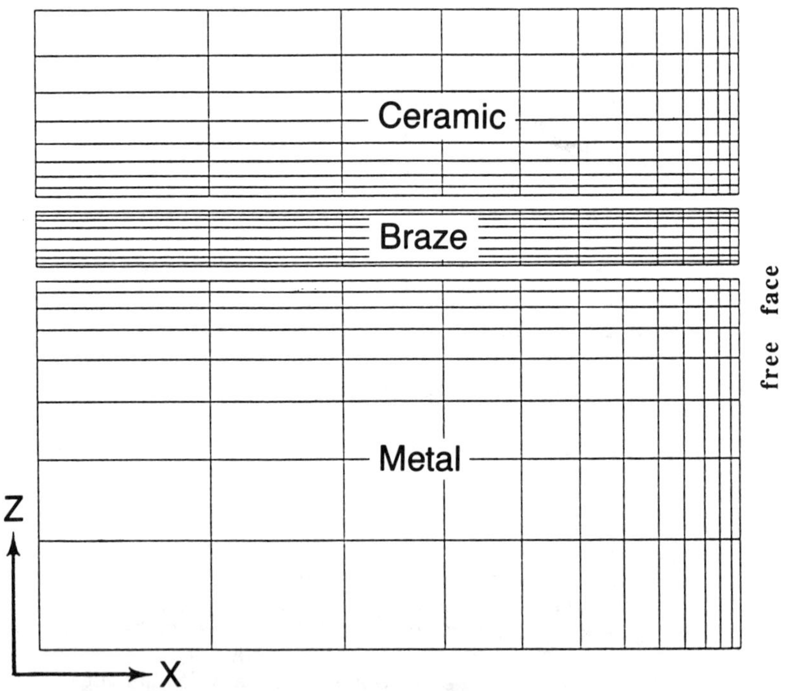

Fig. 3 Standard Finite Element Idealization. *The idealization contains 288 isoparametric quadrilateral elements with midside nodes. Thicknesses of the ceramic, braze and metal are 0.010 in. (254 μm), 0.003 in. (76.2 μm), and 0.020 in. (508 μm), respectively.*

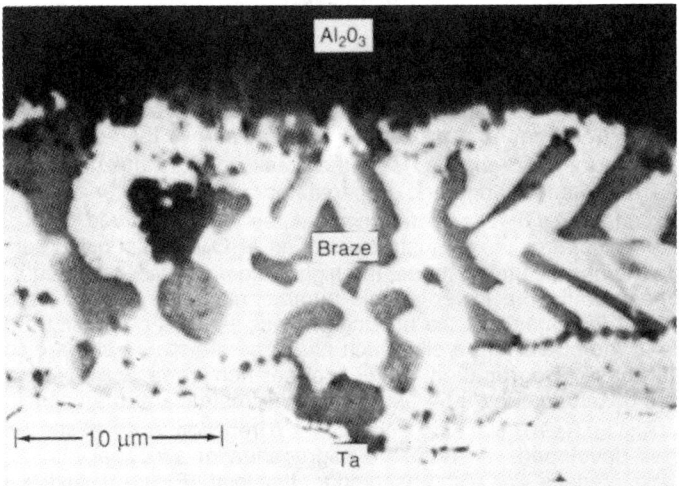

Fig. 4 Interface Region in Brazed Al_2O_3/Incusil/Ti. *In the braze region, grey areas are Cu rich and light areas are Ag rich (4000X).*

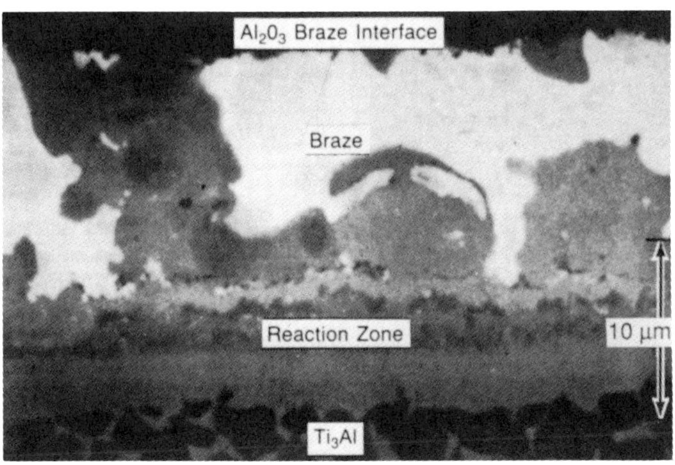

Fig. 5 Backscattered Image of Reaction in Al_2O_3/Incusil/ Ti_3Al (3000X)

3. Results

3.1 Experimental

In the case of SiC/Ti brazed joints, the crack appeared to nucleate in the SiC phase near the free edge and propagate in shear across the specimen parallel to the interface. In the case of Al_2O_3/W, the crack appeared to nucleate at the W-braze interface at the free edge and propagate parallel to the interface. In the cases of Al_2O_3/Ta and Al_2O_3/Ti_3Al, no interfacial failures were observed.

Figure 4 illustrates the microstructure of an Al_2O_3/Ta joint using backscattered electrons. In the backscattered mode, the higher atomic number elements appear as a lighter phase while the lower atomic number elements appear as the darker phase. Figure 4 indicates that the brazing process yields a non-uniform distribution of elements in the braze with a silver-rich phase as the matrix and a discontinuous copper-rich phase. Segregation of the copper-rich zone at the ceramic/braze interface was observed. Figure 5 illustrates the microstructure of an alumina/titanium aluminide (Al_2O_3/Ti_3Al) joint and shows that a reaction zone at the braze-metal interface has developed. In addition, segregation of a copper-rich zone at the ceramic/braze interface was also observed for this joint. Figure 6 illustrates a typical mapping of the elemental distribution at the interface of a SiC/Ti joint and indicates that the silicon has dissolved into the braze alloy leading to Ti and Si-rich regions.

These findings indicate that the presence of reaction zones at the ceramic-braze or metal-braze interface and/or dissolution of elements by the braze alloy leads to a more complicated description of the materials than assumed in the finite element analysis. A more accurate model would take into account the mechanical properties of the reaction zones. However, the finite-element analysis predicts the correct qualitative trends observed experimentally. We conclude that the reaction zones or braze alloy modifications do not appreciably affect the general stress state computed from the more approximate model.

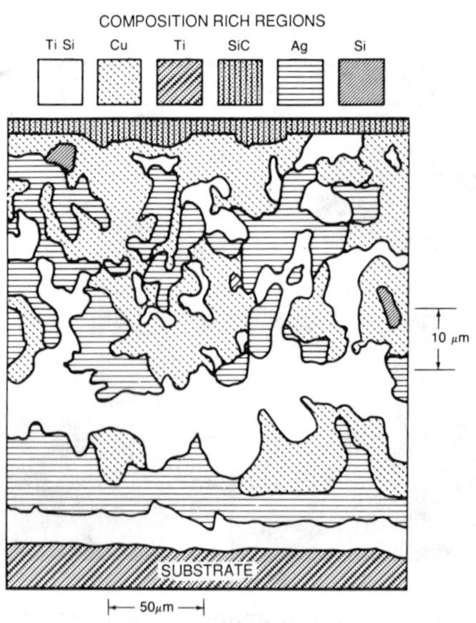

Fig. 6 Elemental Distribution of Braze Alloy of SiC/Incusil/Ti

3.2 Analytical

We first investigate the brazing of SiC and Ti. The deformed shape of a SiC/Incusil/Ti test coupon, after the braze cycle, is shown in Fig. 7. A sharp slope change occurs at the ceramic/braze interface and indicates a region of high strain and corresponding high stress. Figure 8 shows the normal stress component, σ_z, along the free face. The maximum stress (tensile) occurs approximately 5 μm from the SiC/Incusil interface. This suggests crack initiation may occur (mode I) at this point, assuming that the toughness of the interface is as least as high as the toughness of the ceramic. Figure 9 shows the normal stress, σ_z parallel to the SiC/Incusil interface. Note that while crack initiation may occur in tension, a mode I crack cannot propagate as the high tensile stress becomes compressive very close to the free face. Figure 10 shows the shear stress, τ_{xz}, in the ceramic along the SiC/Incusil interface during cooling from the braze temperature to ambient temperature. While crack initiation will occur in tension, the generation of high shear stresses parallel to the interface, as ambient temperature is approached, suggests that crack propagation probably will occur in shear (i.e., mode II). If the toughness of the interface is as high as the ceramic, then a mode II crack may propagate in the ceramic parallel to the interface and a small distance from it. Note that above 300^oC the shear stress remains low due to the low braze yield stress.

Figure 11a shows the residual shear stress at ambient temperature within the brazed joint. In order to investigate whether accurate stresses are obtained near the stress singularity, a "fine" grid was generated and the results shown in Fig. 11b. The shear stress results changed negligibly for the two grids, "standard" and "fine," except within the two elements closest to the free face. Thus, we used the coarser "standard" grid for all FEA. These two elements have a total length of 6 μm in the "fine" grid. In addition, we observed an increase in σ_z at the free face, for the "fine" grid compared to the standard grid results shown in Fig. 6. We concluded that an accurate solution may be obtained even though a singularity exists, and we used the coarser "standard" grid for all our FEA. The ambient temperature curve in Fig. 10 (T=25^oC) matches the SiC/Incusil interface curve in Figs. 11a and 11b everywhere except right at the singularity. In Fig. 11 the shear stress does not decrease to zero at the free face. This anomaly occurs because the assumption of a symmetric stress tensor does not hold at a singularity.

Figure 12 illustrates the effect of the braze work-hardening rate on the shear stress distribution at or near the ceramic/braze interface. As the work-hardening rate increases, the shear stress level is increased. Thus, less work-hardening (i.e., more plastic deformation) decreases the stress level and the likelihood of crack propagation. Next we investigate the effect of joint thickness on the residual stresses developed along the SiC/braze interface. Figure 13a shows the residual shear stress for three different joint thicknesses, assuming that the braze behaves elastically. The assumption of an elastic braze is done for comparative purposes. These results indicate that as the joint thickness increases, the residual shear stress along the ceramic/braze interface increases. Taking into account the plastic behavior of the joint, we note that the thickness effect is less severe (Fig. 13b). Thus the effect of plastic behavior within the joint is to reduce the effect of the joint thickness. For the case considered here, joint thickness variations do not seem to be strongly influential in controlling the residual stresses. Note that τ_{xz} at ambient temperature, is much less than it would be if the braze behaved elastically throughout the cooling process (projecting the results of Fig. 13a by 10 times for a ΔT of 800^oC). The plastic behavior of the braze has the effect of greatly reducing the residual stresses along the SiC/Incusil interface.

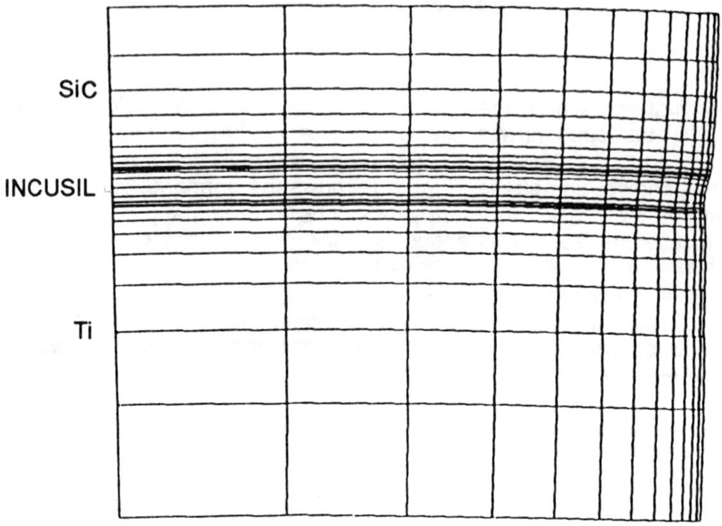

Fig. 7 Deformed Shape Plot of SiC/Incusil/Ti Laboratory Coupon Cooled from Brazing Temperature to Ambient Temperature

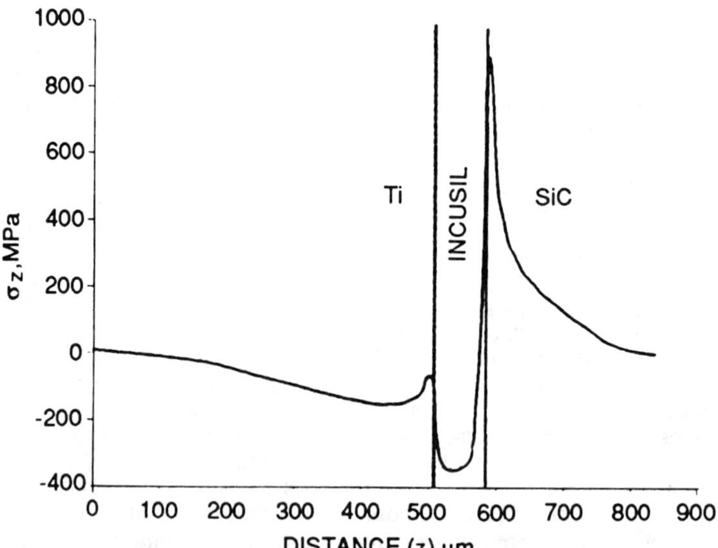

Fig. 8 Residual Normal Stress, σ_z, Along Free Edge of SiC/ Incusil/Ti Laboratory Coupon Cooled to Ambient Temperature

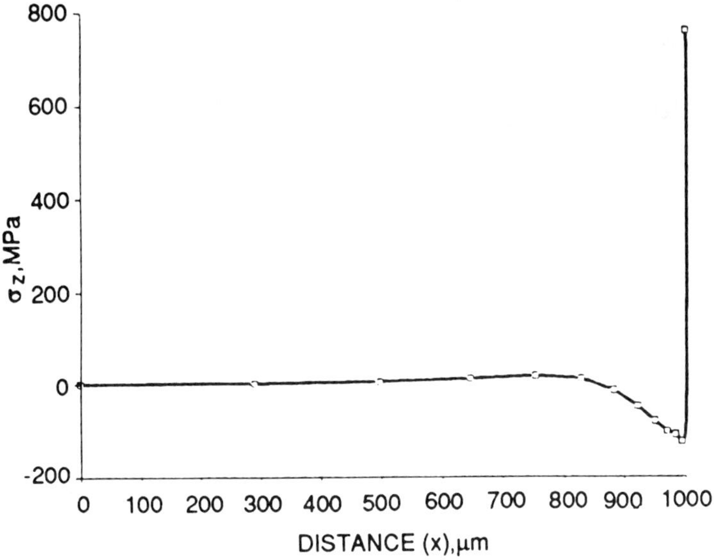

Fig. 9 Normal Stress, σ_z, Parallel to SiC/Incusil Interface, & 5 μm into SiC, at Ambient Temperature

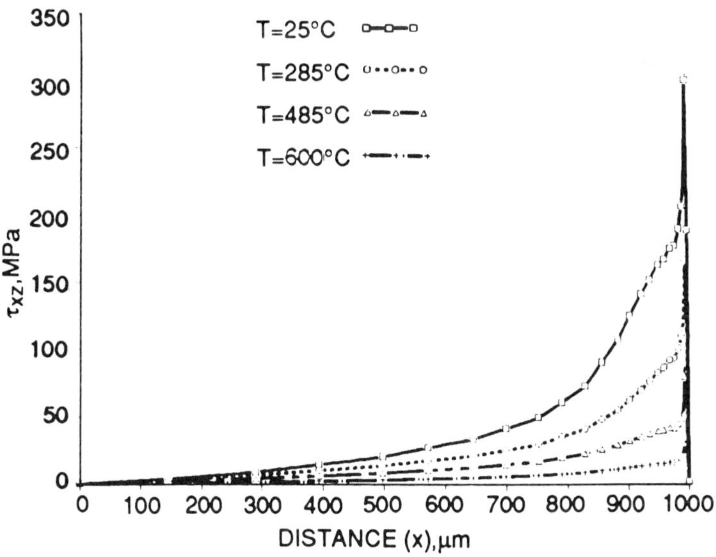

Fig. 10 Shear Stress, τ_{zx}, Along SiC/Incusil Interface, During Cooling Process

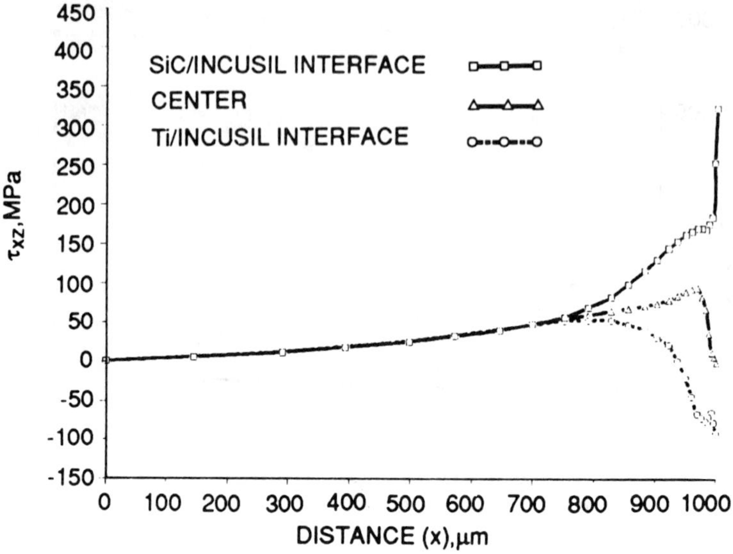

(a) "STANDARD"

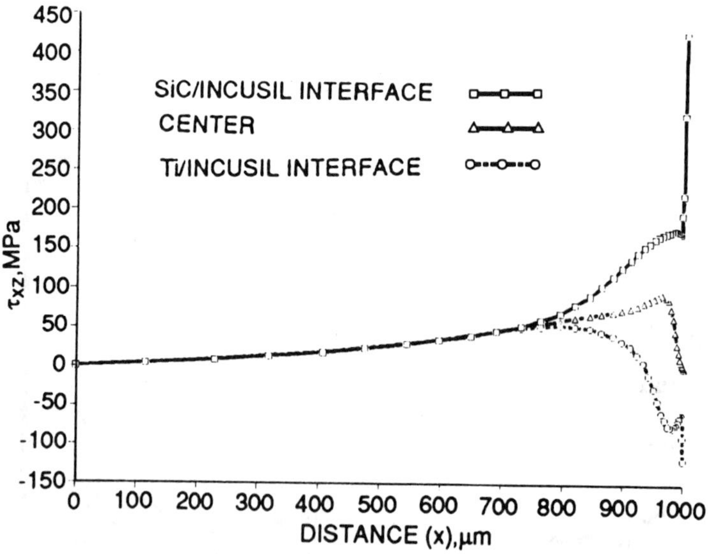

(b) "FINE"

Fig. 11 Residual Shear Stress Within Brazed Joint for "Standard" & "Fine" Finite Element Idealization

One goal of our analysis is to find material combinations in which the residual stresses do not cause failure at the ceramic/braze interface. Lower residual stresses at the ceramic/braze interface may be accompanied by higher residual stresses at the metal/braze interface, where presumably the higher strength of the metal may withstand these stresses without failure. For example, an Al_2O_3/W composite is considered. Figures 14 and 15 illustrate the higher residual shear stress at the metal/braze interface than at the ceramic/braze interface. The peak stresses at the metal/braze interface indicated that failure should occur here due to mode I crack initiation and mode II fracture. This was observed experimentally.

We have chosen this material combination to investigate the practical case of a ceramic tile brazed onto a metal substrate. Figure 16 shows the deformed shape after cooling down to ambient temperature. Note the sharp slope change at the Incusil/W interface. The corresponding increase in peak shear stress at the interface is shown in Fig. 17. Thus, even if brazing is successful demonstrated for a laboratory coupon sample, there is no guarantee that the same materials can be successfully joined in the practical case of joining a ceramic to a metal substrate.

Another case considered is brazing Al_2O_3 to Ta. The shear stress distributions shown in Fig. 18 indicate a more evenly divided stress distribution between the two interfaces, and low enough stress levels to insure a successful bond. Similar conclusions were reached for a brazed Al_2O_3/Ti_3Al joint.

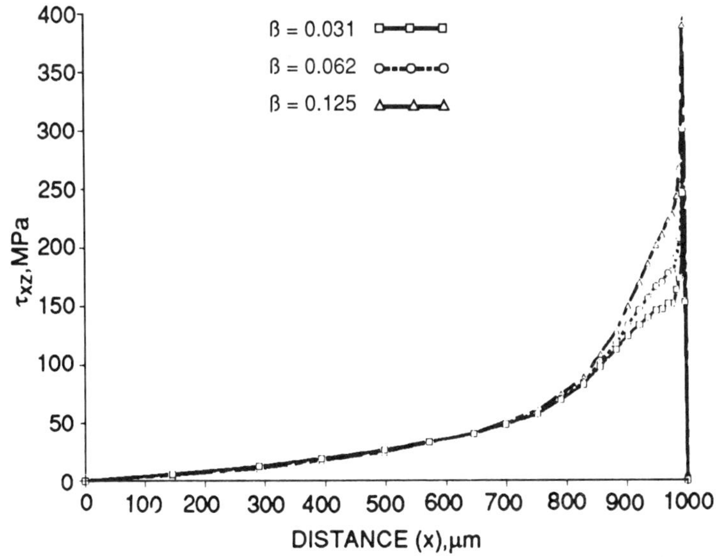

Fig. 12 Residual Shear Stress in Joint for Different Braze Work-hardening Rates, $\beta = E_p / E$; SiC/Incusil/Ti Coupon

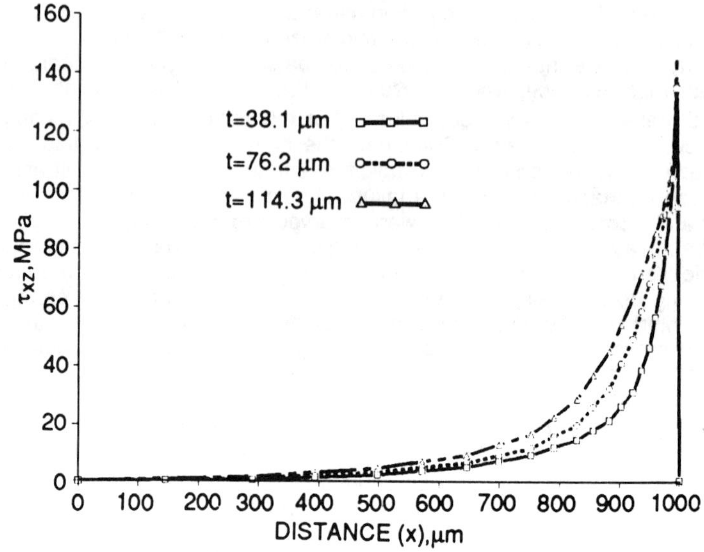

(a) ASSUMING BRAZE BEHAVES ELASTICALLY DURING TEMPERATURE CHANGE OF 80°C

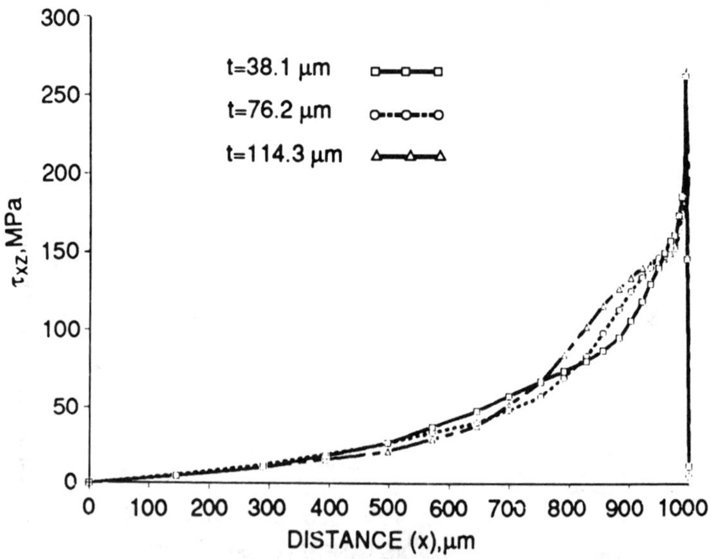

(b) AMBIENT TEMPERATURE, ALLOWING ELASTOPLASTIC BEHAVIOR OF BRAZE DURING COOLING PROCESS

Fig. 13 For SiC/Incusil/Ti Coupon; Shear Stress Along SiC/Incusil Interface for Different Brazed Joint Thicknesses

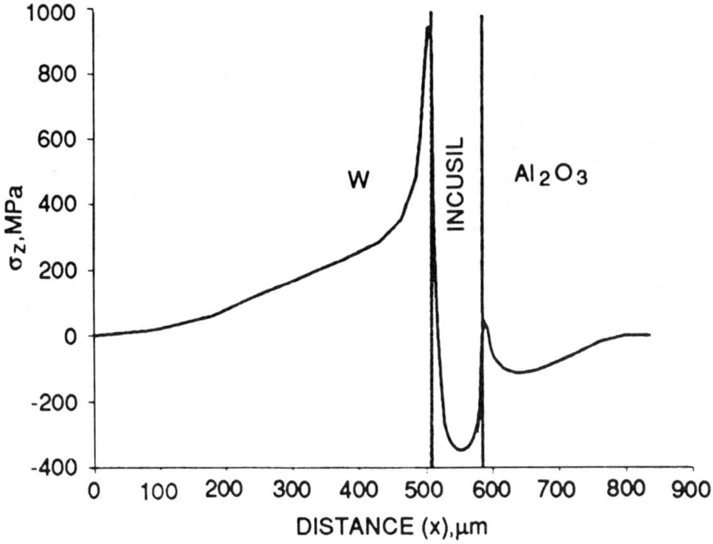

Fig. 14 Residual Normal Stress, σ_z, Along Free Face for Al_2O_3/Incusil/W Coupon, Cooled from Braze Temperature to Ambient Temperature

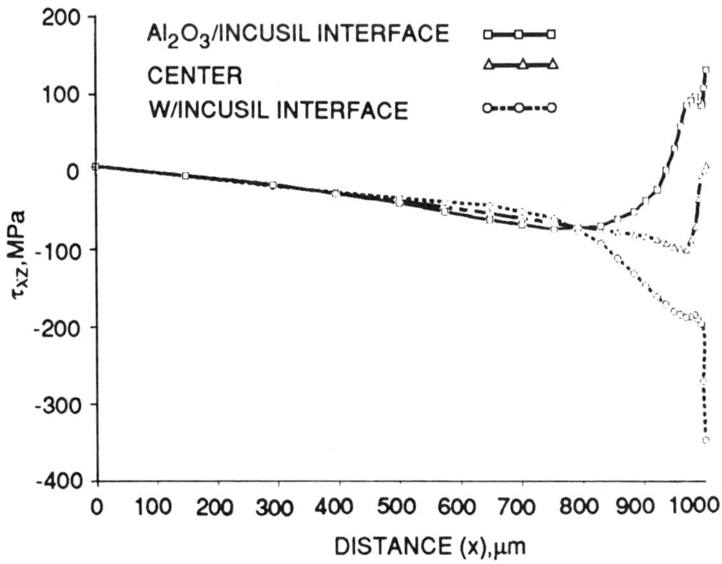

Fig. 15 Residual Shear Stress in Joint for Al_2O_3/Incusil/W Coupon, Cooled from Braze Temperature to Ambient Temperature

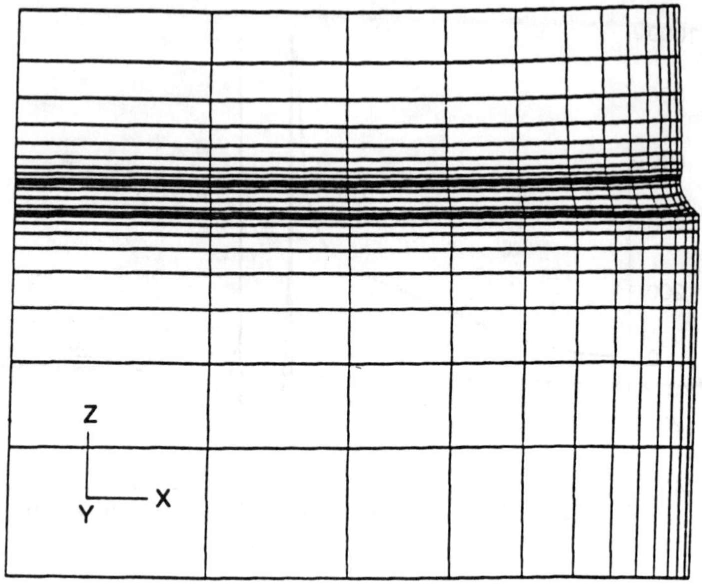

Fig. 16 Deformed Shape for Case of Al_2O_3 Tiles Brazed onto W Substrate

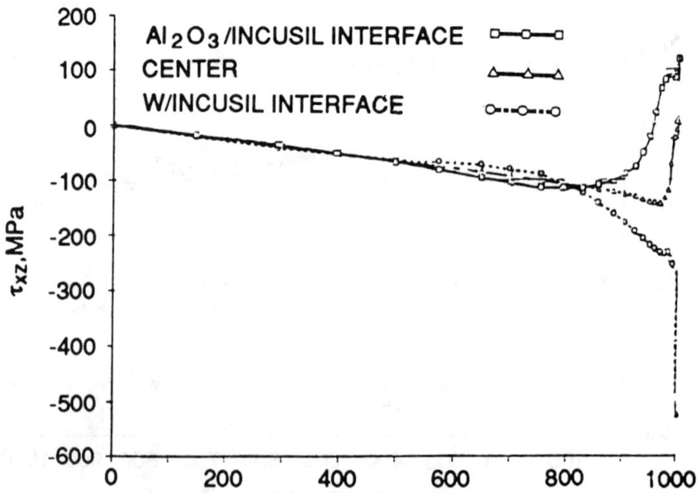

Fig. 17 Residual Shear Stresses in Joint for Case of Al_2O_3 Tiles Brazed onto W Substrate

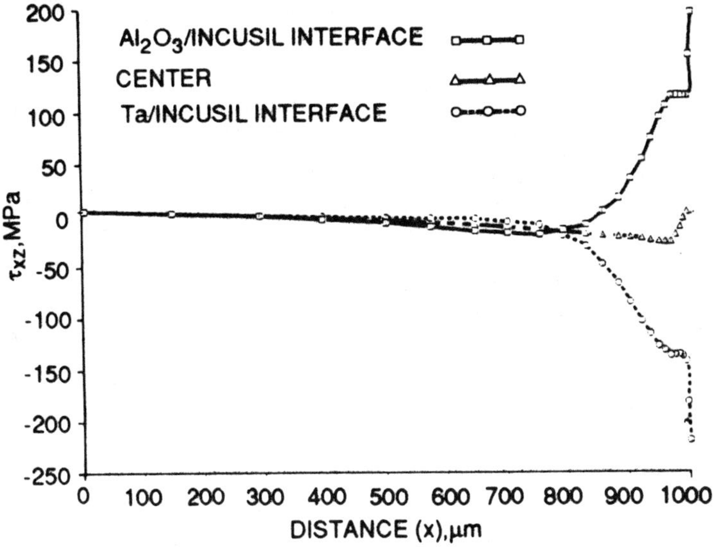

Fig. 18 Residual Shear Stress in Joint for Al_2O_3/Incusil/Ta Test Coupon Cooled from Braze Temperature to Ambient Temperature

Table II illustrates the effect of changing the braze thickness on the shear stress. Column 3 gives the average shear stress obtained from the simplified elastic solution assuming that the edge distance, L, is equal to twice the sum of the ceramic and braze thickness. This is based on the results of the FEA (see Fig. 10a). The elastic FEA (Fig.13a) agrees qualitatively with the closed form solution, i.e., the average shear stress increases with increase in braze thickness, and the peak value obtained from the FEA is approximately twice the average value obtained from the simplified solution. However, with further changes to room temperature, the braze deforms plastically and the shear stress profiles for the differing braze thicknesses tend to converge (Fig. 13b). Thus, for the case considered here, the braze thickness does not seem to be an overly influential parameter for controlling the residual stresses.

Table II. Residual Shear Stress at SiC/Incusil Interface Due to Cooling from 825°C

t_{braze}(μm)	L(μm)	Closed-Form Sol. τ, MPa @ΔT=800°C	FEA (elastic) τ, MPa @ΔT=80°C	FEA (elastoplastic) τ,MPa@ΔT=800°C
114.3	728.6	72.5	75	175
76.2	652.4	66.1	71	185
38.1	576.2	58.2	65	201

where, L=2 x (t_{braze} + $t_{ceramic}$)

4. Discussion

We confirmed the conclusions of Reference 11 for a thermal problem involving a multilayered material. We demonstrated that accurate stress results can be obtained near a stress singularity, namely where the intersection of dissimilar materials approaches a free edge.

Residual stresses developed during the cooling process from the braze temperature to ambient temperature may be high enough to cause failure during the brazing process either at or near the ceramic/braze interface or metal/braze interface. In all cases considered, the residual stresses appear high enough that they must be considered in any analysis of properties or strength due to subsequent mechanical or thermal loading. Including in our analysis the temperature-dependent braze stress-strain response gives us an accurate description of the stress response during the cooling process. This information may lead to an efficient stress relaxing procedure in which, during the cooling process, the temperature is held at an intermediate level (at which the residual stresses are high, but not catastrophic) to allow for a stress relaxation through creep. For the case of SiC/Incusil/Ti, this would occur at approximately 300°C.

Experimental findings for SiC to Ti are consistent with the analytical findings, in that crack initiation took place in the SiC phase near a free edge in the vicinity of the SiC/Incusil interface and propagated parallel to the interface. Also, for Al_2O_3 brazed to W, failure occurred at the metal/braze interface, while, for Al_2O_3 brazed to Ta or Ti_3Al, the brazing process was successful.

From our findings we conclude that the mismatch of CTE between the ceramic and metal leads to detrimental residual stresses. As the coefficients approach each other, Al_2O_3/Ta for example, the residual stresses decrease. When the ceramic has a higher CTE than the metal the critical area appears to be near the ceramic/braze interface, while when the CTE of the metal is higher, the critical area appears to be at the metal/braze interface. Patten, et al.,[2] found similarly that when brazing ceramics to metals having a significantly larger CTE, failure normally occurred in the ceramic at or near the free face of the ceramic/braze interface. We conclude, although with limited evidence, that if the metal is tougher than the ceramic, it is better to have a lower CTE in the metal than in the ceramic. This should be explored further and may prove to be very useful.

Experimental findings indicate a reaction zone near the joint interfaces. These zones should be more clearly identified and taken into account in the analytical predictions of the residual stresses developed in the joint. This could be done by introducing, within a FEA, very thin reaction layers at the joint interfaces.

5. Conclusions

- Accurate stress results can be obtained for a thermal problem involving a multilayered material, where a stress singularity may exist
- Failure at or near an interface may occur during the brazing process. This failure often occurs as a mode I crack initiation and a mode II crack propagation
- Generally, the residual stresses developed during the cooling process will be sufficiently high to influence the strength and properties of the brazed joint during subsequent mechanical or thermal loadings, and therefore must be accounted for
- Increase in the braze work-hardening rate increases the peak shear-stress levels
- Increasing the brazed joint thickness generally leads to higher peak stresses, but plastic behavior within the joint reduces this effect

• Higher stresses develop at the metal/braze interface when the metal is constrained to remain planar as in the case of ceramic tiles brazed onto a metal substrate. Thus, even if brazing is successfully demonstrated for a laboratory coupon, there is no guarantee that the same materials can be successfully joined in the case of joining a ceramic to a metal substrate

• Qualitative correlation is found between analytic results and experimental findings regarding the feasibility of brazing the ceramic/metal combinations considered here

• Exploring joints in which the metal has a lower CTE may prove significant.

• The formation of reaction zones at the joint interfaces should be further investigated to identify them and their influence on the joint residual stresses.

6. References

1. P. Charreyron, D. Patten, and B. Miller, "Modelling of Ceramic To Metal Brazed Joints," Ceram. Engng. Sci. Proc. - Intl. Forum on Structural Ceramics Joining, 10 (11-12) (Westerville, OH: American Ceramic Society, 1989), 1801-1824.

2. D. Patten, M. Torti, and, P. Charreyon, "Mechanical Behavior of Ceramic-Metal Brazed Joints," Ceram. Engng. Sci. Proc. - Intl. Forum on Structural Ceramics Joining, 10 (11-12) (Westerville, OH: American Ceramic Society, 1989), 1866-1878.

3. H. Cao, M. Thouless, and A. Evans, "Residual Stresses and Cracking in Brittle Solids Bonded with a Thin Ductile Layer," Acta Metallurgica, 36 (8) (1988), 2037-2046.

4. A. Evans, M. Lu, S. Schmauder, and M. Rühle, M., "Some Aspects of the Mechanical Strength of Ceramic Metal Bonded Systems," Acta Metallurgica, 34 (8) (1986), 1643-1655.

5. B. Dalgleish, M. Lu, , and A. Evans, "The Strength of Ceramics Bonded with Metals," Acta Metallurgica, 36 (8), (1988), 2019-2035.

6. W. Kingery, Introduction to Ceramics, (New York, NY: John Wiley & Sons, Inc. 1960), 201-202.

7. A. Levy, "High-Temperature Inelastic Analysis," Computers and Structures, 13 (1981), 249-256.

8. A. Levy and A. Pifko, "On Computational Strategies for Problems Involving Plasticity and Creep," International Journal for Numerical Methods in Engineering, 17 (1981), 747-771.

9. Y. Veda, and T. Yamakawa, "Thermal Nonlinear Behavior of Structures," Advances in Computational Methods is Structural Mechanics and Design, Proc. of 2nd US-Japan Seminar on Matrix Methods of Structural Analysis and Design, August 1972), 375-392.

10. Y. Fung, Foundations of Solid Mechanics, (New Jersey: Prentice-Hall, 1965), 67.

11. J. Whitcomb, J. Raju, and J. Goree, "Reliability of the Finite Element Method for Calculating Free Edge Stresses in Composite Laminates," Computers and Structures, 15 (1) (1982), 23-37.

PEEL ADHESION BOND STRENGTH OF DIRECT BONDED COPPER-ALUMINA

AS AFFECTED BY ALUMINA SINTERING AIDS

By John E. Holowczak, Victor A. Greenhut and
Daniel J. Shanefield

Center for Ceramics Research,
Rutgers, The State University of New Jersey,
Piscataway, NJ 08855-0909

Abstract

The gas-metal eutectic (direct bond) method was used to attach copper foil to sintered high alumina ceramics which had different sintering aid compositions in the magnesia-calcia-silica system. The highest average copper-alumina peel adhesion strength, 205 N/cm, was observed for alumina which contained 0.2% magnesia and 0.2% calcia. The lowest adhesion strength, 103 N/cm, was observed for copper bonded to 95% alumina which contained magnesia, calcia and silica additions. Statistical matrix experiments showed that alumina containing calcium and silicon oxides had significantly lower copper bond strength.

Metal - Ceramic Joining
Edited by P. Kumar and V. A. Greenhut
The Minerals, Metals & Materials Society, 1991

Introduction

In the past, considerable attention has been paid to both atmosphere and composition of the metallic phase(s) and their subsequent effects on ceramic to metal bond strength.[1-3] Relatively few investigators[4] have studied the effects of ceramic composition on bond strength. In this study, the composition of the minor addition(s) commonly used as sintering aids for high alumina was systematically varied using a designed experiment. The effect of these changes on adhesion of copper were evaluated using peel testing.

The use of the gas-metal eutectic (direct bond) method for joining metals to ceramics has attracted interest for both electronic and structural applications.[5-7] This method allows a simple, elemental metal (as well as some alloys) to be bonded to a ceramic without the use of applied pressure. The technique was originally developed for the electronic packaging industry, but has recently received attention as a method for bonding copper and nickel to ceramics for structural applications. Recent technological demands have provided a reason for renewed interest in this area; however, many of the advances remain proprietary.

Gas-metal eutectic bonds are formed by heating a copper/ceramic assembly in an inert atmosphere in the presence of small amounts of oxygen. The oxygen may be added to the atmosphere, or, alternatively, the oxygen may be provided by pre-oxidizing the copper by chemical or thermal means. When the copper reaches a temperature of 1065°C, a eutectic copper-copper oxide skin forms on the surface of the metal. This liquid skin wets most oxide ceramics, and upon cooling, will segregate to form copper with precipitated copper oxide that adheres strongly to the ceramic. The copper-oxygen phase diagram is illustrated in Fig. 1, and was constructed using data from Refs. 8 and 9.

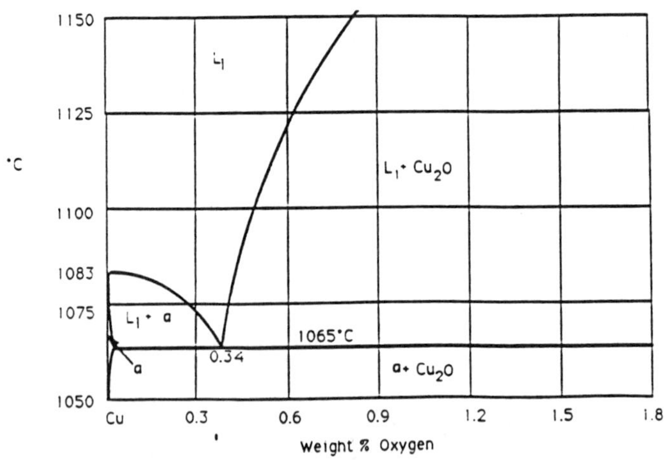

Figure 1 - The copper-oxygen binary phase diagram. The diagram was constructed using data from references 8 and 9.

Experimental

A 3^2 factorial matrix experiment was designed as illustrated in Fig. 2. The oxides listed in the squares of the matrix were those that were added as sintering aids. The bulk compositions are listed in the same format in Fig. 3. Alumina was processed by mixing Alcoa A-16SG with the appropriate amounts of additives by ball milling in alcohol for 10 h. The slurries were then dried, dry pressed at 55 MPa, and fired to 1650°C for 30 min. Fired densities for all components were greater than 97.5% of theoretical by the Archimedes method. Several compositions were checked subsequent to sintering using wet chemical analysis and were found to approximate the starting composition.

Surface species concentrations for each of the alumina compositions were characterized using XPS and are displayed in Fig. 4. As expected, most of the sintering aids were present on the surfaces in amounts greater than the bulk chemistries would indicate. Surface topographies of the various alumina compositions were similar as measured using profilometry, but only small variations were found.

Gas-metal eutectic bonding was carried out using thermal pre-oxidation of the copper foil to provide the required oxygen. Electrolytic tough pitch copper foil 250 um (0.01 in.) thick was cleaned by etching in 15% nitric acid followed by rinsing in distilled, de-ionized water. Foils were then heated in air at 100°C for 24 hours to form a dark, orange colored oxide layer. Alumina samples were pre-fired in air to 650°C for one hour immediately prior to insertion in the bonding furnace. Copper foils were placed on top of the alumina samples and then the assembly was placed in a tube furnace under a flowing nitrogen atmosphere. The copper-alumina assemblies were then heated above the eutectic temperature to approximately 1072°C and held for 180 sec.

	Trace	MgO	MgO, SiO_2
Trace	Trace	MgO	MgO SiO_2
CaO	CaO	CaO MgO	MgO CaO Low SiO_2
CaO, SiO_2	CaO SiO_2	MgO CaO Low SiO_2	CaO MgO High SiO_2

Figure 2 - Factorial matrix experimental design. The oxides listed in the squares of the matrix were those used as sintering aids.

Al_2O_3 99.9%	MgO 0.2% Al_2O_3 99.7%	MgO 0.2% SiO_2 0.2% Al_2O_3 99.5%
CaO 0.2% Al_2O_3 99.8%	MgO 0.2% CaO 0.2% Al_2O_3 99.5%	MgO 0.5% CaO 0.5% SiO_2 0.6% Al_2O_3 98.4%
CaO 0.8% SiO_2 0.8% Al_2O_3 98.3%	MgO 0.5% CaO 0.5% SiO_2 0.6% Al_2O_3 98.4%	MgO 1.5% CaO 1.5% SiO 2.0% Al_2O_3 95%

Figure 3 - Bulk compositions of alumina ceramics produced for this study.

O 66.9% Al 31.8% Mg 0.6% Ca 0.3% Si 0.0%	O 68.6% Al 29.5% Mg 1.3% Ca 0.1% Si 0.0%	O 65.2% Al 32.5% Mg 1.2% Ca 0.3% Si 0.6%
O 66.6% Al 31.6% Mg 0.6% Ca 1.1% Si 0.0%	O 65.9% Al 30.4% Mg 2.1% Ca 1.3% Si 0.0%	O 69.5% Al 26.3% Mg 0.0% Ca 1.4% Si 2.9%
O ... Al ... Mg ... Ca ... Si ...	O 69.5% Al 26.3% Mg 0.0% Ca 1.4% Si 2.9%	O 63.4% Al 23.3% Mg 0.2% Ca 1.6% Si 11.8%

All samples had 0 to 0.35% Na on surface.

Figure 4 - Surface species of alumina compositions determined by XPS. Amounts of ubiquitous carbon at the surfaces were excluded.

After bonding, the strengths of the joins were tested using an improved peel adhesion technique similar to the method employed by Mizuhara and Mally.[10] Peel Strips were 0.635 cm wide and were tested at a crosshead speed of 0.27 cm/min. For a detailed description of the peel adhesion technique and critical factors associated refer to the Appendix of this article.

Multilayer copper-alumina samples were prepared using a commercial 96% alumina. These were sectioned perpendicular to the interfaces and used to prepare transmission electron microscopy samples. Samples were thinned to approximately 150 um using diamond grinding and cut into 3 mm discs using a diamond core drill. Samples were then mechanically dimpled to a center thickness of less than 20 um, followed by ion milling at a low angle (less than 7°) using argon.

Results and Discussion

The results of the peel adhesion tests are shown in matrix format in Fig. 5. An analysis of variance (ANOVA) showed that there were statistical differences and interaction effects in the matrix at a 99% confidence level. It was also observed that by changing the alumina composition the copper bond strength could be increased from 103 N/cm for a 95% alumina composition to a high of 205 N/cm for 99.5% alumina which contained small amounts of magnesia and calcia. Commercial 96% alumina showed bond strengths similar to the 95% alumina prepared for this study: 113 N/cm. The specific composition of both materials was quite similar. This implies that adjusting the specific compositions of alumina ceramics can indeed enhance bonding.

In Fig. 6, the upper left corner of the matrix was analyzed separately using ANOVA. Although no statistical differences in bond strength were produced by adding only calcia or only magnesia to the alumina, there was a significant interaction effect within a 99% confidence interval. This effect indicated an increase in bond strength produced by small additions of both calcia and magnesia. The reason for the increase in bond strength is as yet unclear. The effect could possibly be either synergistic or due to increased total levels of Group II elements being present.

	Trace	MgO	MgO, SiO_2
Trace	Tr. 163 ± 23	Mg 147 ± 26	Mg,Si 173 ± 12
CaO	Ca 144 ± 33	Ca,Mg 205 ± 14	Mg,Ca,lo Si 172 ± 23
CaO, SiO_2	Ca,Si 159 ± 14	Mg,Ca,lo Si 172 ± 23	Mg,Ca,hi Si 103 ± 11

Figure 5 - Average peel adhesion values in N/cm for the matrix experiment. An ANOVA showed significant differences between the compositions in the matrix within a 99% confidence interval.

Trace 163 ± 23	MgO 147 ± 26
CaO 144 ± 33	MgO,CaO 205 ± 14

Figure 6 - Upper left corner of the matrix experiment. A significant interaction effect was observed for calcia and magnesia additions, with greater bond strength when both were present.

Figure 7 illustrates the effect of increasing surface silica content for aluminas which also contained both magnesia and calcia. As can be observed, increased levels of silicon at the surface appear to reduce copper bond strength. This could possibly be caused by the increased glassy phase inhibiting or preventing the formation of copper aluminate as an interfacial bridging oxide.

The entire 3^2 matrix of peel strength results were analyzed using Scheffe's analysis to test for differences between columns and rows. The analysis showed that the bottom row of the matrix (calcium silicate containing aluminas) had significantly lower copper bond strengths at a 95% confidence interval, as compared to the rest of the matrix.

Transmission electron microscopy was used earlier in this study to examine copper-alumina interfaces using a commercial alumina which contained both calcium and silica. A micrograph of the interface is shown in Figure 8. The uppermost layer is primarily copper, while the bottom (dark) layer is aluminum oxide as revealed by energy dispersive spectroscopy (EDS). An EDS analysis of the center band is shown in Fig. 9, and reveals that this layer contains significant amounts of copper, aluminum and oxygen. Phosphorous was believed to have diffused from the bulk copper. Phosphorous is commonly used as a de-oxidizing agent in copper refining. Significant amounts of calcium, as well as silicon, were also detected in this interfacial layer.

The TEM-EDS evidence clearly shows that both calcium and silicon will effect copper-alumina bonding. Strength data discussed above indicate that, when

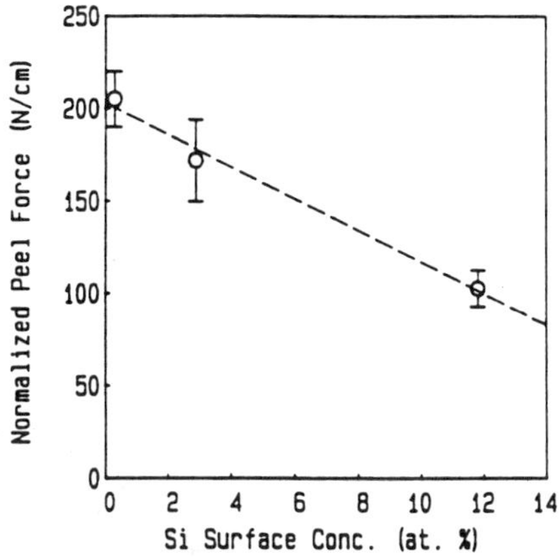

Figure 7 - Effect of increasing surface silicon on copper bond strengths for alumina which contained calcia and magnesia.

alumina containing calcium and silicon oxides is used as a substrate for copper bonding, lower strengths will result. It is speculated that a bridging oxide other than the commonly observed copper aluminate (found in copper/alumina systems) is formed. This phase might be copper calcium oxide or a copper calcium silicate.

Conclusion

At a 99% confidence interval, alumina composition will effect gas-metal eutectic copper bond adherence. Increasing silica content appears to reduce bond strength, particularly when calcium is also present in the alumina. TEM observations show that calcium and silicon are present at increased levels directly at the copper-alumina interface. Lower bond strengths could be due to the formation of interfacial phases other than copper aluminate.

Acknowledgments

The authors would like to thank Valeria Greco and Amy Rusby for their help throughout this work. The authors are grateful to Madeline Roepche for performing TEM observations. This work was sponsored by the Center for Ceramics Research, Rutgers University. The support of the New Jersey Commission on Science and Technology and our industrial sponsors is gratefully acknowledged.

Figure 8 - Transmission electron micrograph of eutectic-alumina interface. Top layer is primarily copper. Bottom (dark) layer is alumina. The central band is an interfacial reaction layer.

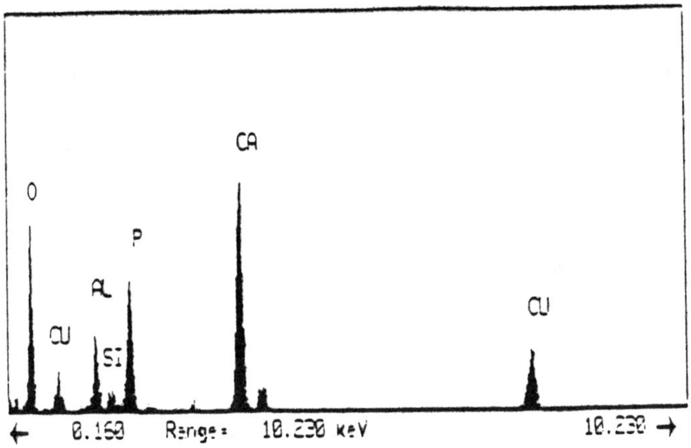

Figure 9 - TEM-EDS analysis of center layer in Figure 8.

Appendix - Peel Adhesion Testing of Ceramic-Metal Bonds

There are several methods available for evaluating the strength of ceramic to metal bonds. Standard tensile and simple lap shear tests may be used when the ceramic to metal join has low to moderate strength. However, when the tensile strength of the bond exceeds 70 to 105 MPa, the failure of the test specimen will normally occur within the ceramic when these tests are used.

Thus as ceramic to metal bond strengths increase, accurate evaluation of interfacial bond strength becomes difficult. Some researchers have utilized four point bend tests to evaluate strong joints. While this method will promote interface failure, fracture within the ceramic or the metal is common, as well as mixed mode fracture. Another disadvantage of many of the tensile, shear and bend tests is the complex, expensive ceramic machining required to fabricate a test sample. Small flaws introduced during machining may strongly influence or overwhelm inherent bond strength results. Setup time is usually time consuming and test fixturing may be complex.

Peel adhesion testing is commonly used in the printed circuit board industry for measuring the adherence of metal foils to plastics. This test has, more recently, been applied to ceramic-metal bonds. Mizuhara and coworkers have used the peel adhesion test successfully for measuring bond strength of active metal brazes. They found that the peel adhesion test was more sensitive to changes in braze composition than tensile or lap shear tests. While peel testing does simulate a common failure mode seen in service, a drawback of this test is that data obtained is not directly applicable to design due to the complex loading.

An important aspect of peel adhesion testing is the width of the adherend strip employed (copper in the present case). Mizuhara notes that if strip widths are too narrow, edge effects will overwhelm the test. If strip widths are too large, thermal expansion mismatch stresses will contribute to interface failure. The mechanics of the test are complex, making it difficult to determine the appropriate strip width. One might expect that thermal expansion and elastic modulus of both metal and ceramic, as well as the thickness and yield strength of the metal will both effect the point at which edge effects will overwhelm the test.

The test apparatus is diagramed in Figs. 10 and 11. The apparatus is affixed to an Instron frame. The chain allows the copper to be peeled at a nearly constant 90° angle from the ceramic. The ceramic sample is affixed to the baseplate of the testing machine. The chain is connected to a load cell, and contains a ball and socket swivel to minimize rotational moments. The length of the chain/swivel assembly should be at least 0.5 m. The chain/swivel is attached to the copper (or other metal) strip. The metal is then peeled from the sample at a constant crosshead speed. Load is measured throughout the test and is plotted as load vs. displacement.

The load vs. displacement curve will often exhibit a peak during the initiation of peeling. This value is ignored, as it is not representative of the constant peeling force. Blisters and other defects will be observed as a momentary decrease in peel load. These values are ignored for the purposes of evaluating the innate bond strength, however, the information is useful as a quality control measure. The average peel force for the remainder of the test is determined and converted into N/cm.

In order to determine the appropriate strip width for the study discussed in this paper, a limited series of tests using different copper strip widths were performed. The strips were bonded to optical grade single crystal sapphire using methods described above.

The results of the tests using various strip widths are shown in Fig. 12. As can be seen in Fig. 12, the strength observed is approximately constant until strip widths are less than 5.5 mm. At high strip widths (10 mm), thermal expansion mismatch stresses may be contributing to failure. Based on these results, a constant strip width of 6.35 mm wide was chosen for the work described here.

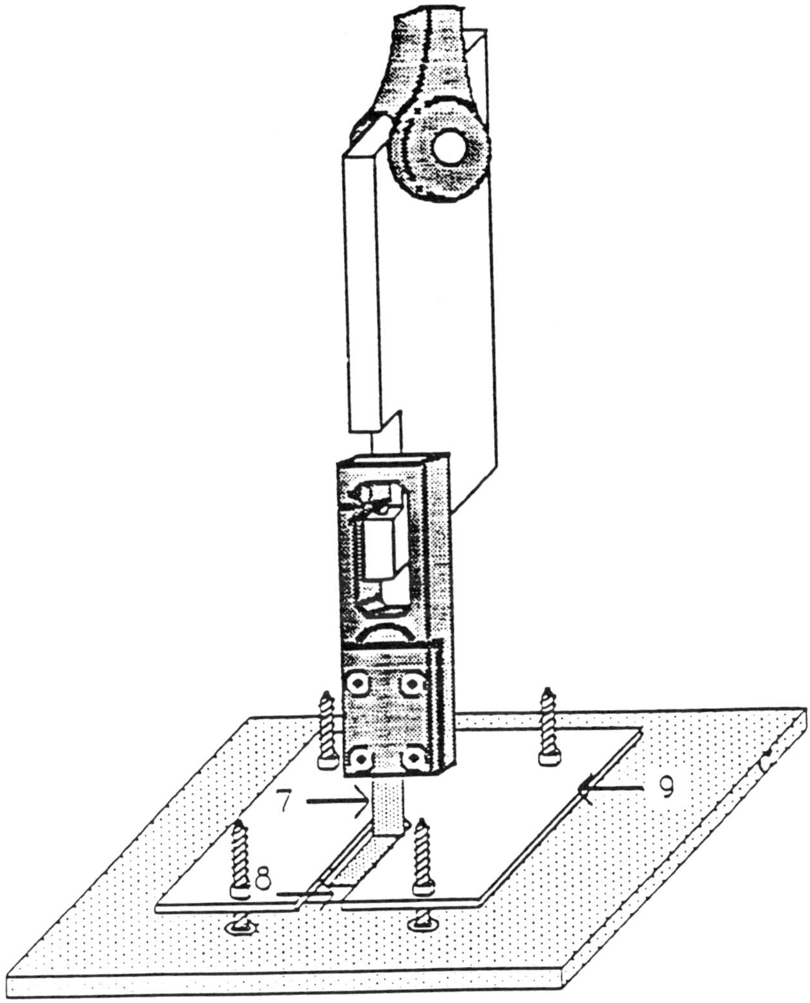

Figure 10 - Peel adhesion test apparatus, 3/4 view.

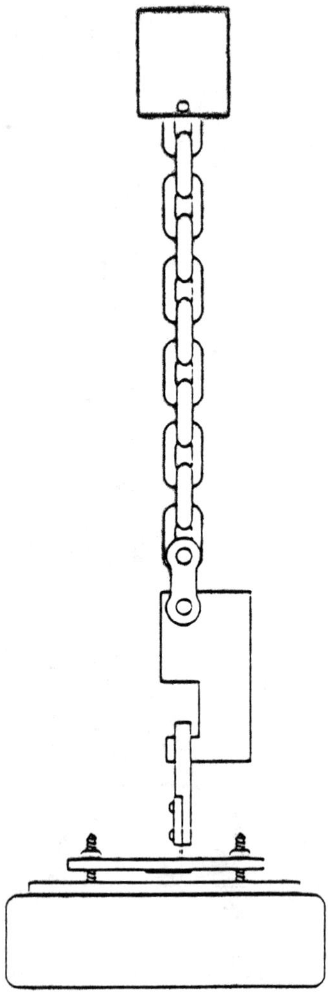

Figure 11 - Peel adhesion test apparatus, side view.

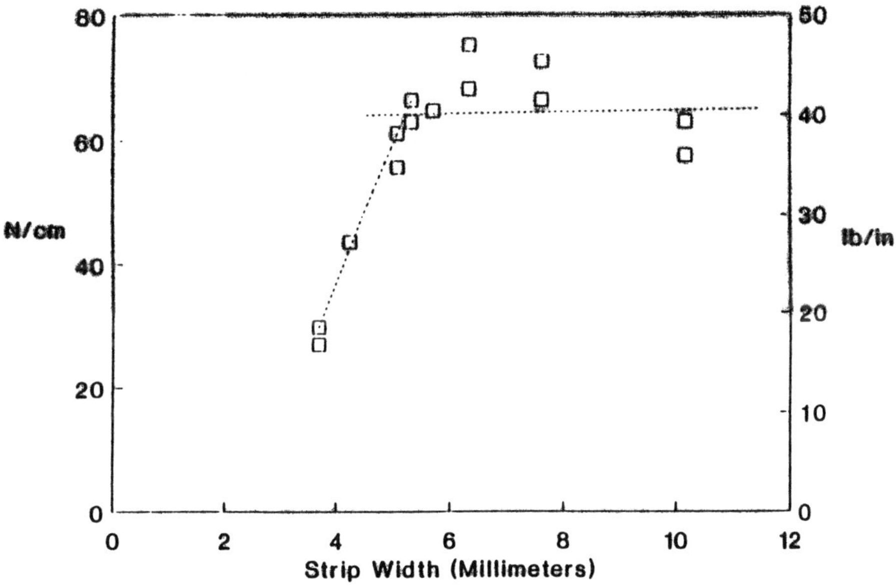

Figure 12 - Peel adhesion strength vs. strip width.

References

1. L. H. Lafarge, "Applications of Ceramic Sections in High Temperature Pulsed Klystrons," Am. Cer. Soc. Bull., 35, [3], 117-132 (1956)

2. J. Schmidt and J. L. Carter, "Using Nitrogen Based Atmospheres for Glass to Metal Sealing," Metal Progress, July, 1985, pp. 29-31.

3. R. M. Fulrath and E. L. Hollar, "Manganese Glass-Molybdenum Metallizing Ceramics," Am. Cer. Soc. Bull., 47, [5], 493-501 (1968).

4. J. R. Floyd, "Effect of Composition and Crystal Size of Alumina Ceramics on Metal to Ceramic Bond Strength," Am. Cer. Soc. Bull., 42, [2], 65-70 (1963).

5. J. F. Burgess, et al., "Hybrid Packages by the Direct Bond Copper Process," Solid State Science and Technology, May, 1975, pp. 40-44.

6. Wittmer, et al., "Mechanical Properties of Liquid Phase Bonded Copper-Ceramic Substrates," J. Am. Cer. Soc., 65, [3], 149-152 (1982).

7. H. Keser, et al., "Some Properties and Applications of Liquid Phase Bonding Between Copper and Ceramics," DVS Berichte, 66, 33-36, 1981.

8. J. F. Burgess, C. A. Neugebaur, and G. Flanagan, "The Direct Bonding of Metals to Ceramics by the Gas-Metal Eutectic Method," J. Electrochemical Soc., 1974.

9. J. P. Neumann, et al., "The Cu-O Binary," Metal Progress, Sept. 1985, pp. 85-88.

10. H. Mizuhara and K. Mally, "Ceramic to Metal Joining with Active Metal Filler Metal," Welding J. [10], 27-32 (1985).

JOINING OF PARTIALLY STABILIZED ZIRCONIA TO

NODULAR CAST IRON

Eunsung Park and James D. Cawley

Department of Materials Science and Engineering
The Ohio State University
Columbus, Ohio 43210

Daniel Hauser

Edison Welding Institute
Columbus, Ohio 43212

Abstract

Mg-partially stabilized zirconia (PSZ) to nodular cast iron joints were fabricated using Ti-bearing brazing systems. The active-substrate and the active-filler-metal processes were studied. Voids were found in the active-substrate joints and are believed to be the result of Ti oxidation prior to the melting of the filler metal. Three possible sources for oxygen are discussed (the filler metal, the PSZ, and the atmosphere). Calculations using equilibrium thermodynamics suggest that both the PSZ and atmosphere are significant sources. Alumina joints also were prepared and the results are compared those obtained using PSZ.

This work was performed under ORNL subcontract 86X-SBO46C, the prime sponsor being the Department of Energy.

Metal - Ceramic Joining
Edited by P. Kumar and V. A. Greenhut
The Minerals, Metals & Materials Society, 1991

Introduction

Joining of ceramics to metals has recently been the subject of extensive study for structural and electronic applications. Brazing is widely used because complex assemblies can be joined fairly easily and the fabrication temperatures are low enough to avoid degradation of either the ceramic or metal parts. Until recently however, ceramic surfaces were treated prior to brazing to assure wetting by the molten filler metal. Numerous experimental studies on the wetting behavior of liquid metals on ceramic surfaces have shown that wetting and spreading can be enhanced by employing an active metal such as Ti[1–4].

Titanium can be introduced either as a coated layer on the ceramic surface or as an element dissolved in the filler metal. The brazing process based on the former is called the active–substrate process and the latter is referred to as the active–filler-metal process. Both processes rely on the formation of a titanium–containing layer at the interface between the brazing alloy and the ceramic. The only difference is how titanium is initially distributed in the system.

Regardless of the brazing process used, titanium tends to react with most anions of ceramic materials (O, C, N) producing Ti–rich reaction layers (TiO, TiC, TiN) at the ceramic–filler metal interface. In the case of SiC and Si_3N_4, Ti also reacts with Si to form titanium silicides.[5,6] These Ti–containing reaction layers are known to play a very important role in producing strong joints. Several studies show a correlation between the formation of these layers and brazing parameters, such as filler metal composition,[7] brazing temperature,[5,8] and brazing time.[5] However, these parameters are not sufficient to define the properties of brazed joints. Additional factors such as the initial distribution of the titanium or other active elements, the brazing atmosphere, and the transport properties of ceramic materials also must be considered. The objective of this work was, therefore, to investigate how these parameters affect the brazing results by comparing two different brazing processes.

Experimental

These experiments were divided into two different groups in terms of the size of specimens (large and small contact area joining) and two groups in terms of the brazing process (the active– substrate and the active–filler-metal processes).

In large contact area joining, Mg–partially stabilized zirconia (Mg–PSZ)* was brazed to nodular cast iron (NCI)# using two different brazing processes (the active–substrate and the active–filler-metal). The Mg–PSZ specimens were 50.8 mm dia. x 19.05 mm cylinders and the NCI specimens were 50.8 mm x 50.8 mm x 19.05 mm parallelepipeds. The area of contact during joining was 20 cm^2. In small contact area joining, 2.54 mm x 5.08 mm x 19.05 mm parallelepipeds of Mg–PSZ were brazed to NCI by the active-substrate process. The contact area in this case was 0.97 cm^2.

* Nilcra, Elmhurst, IL
# Sandia Lab., Albuquerque, NM

The brazing filler metals used were Braze 604 (Ag-30Cu-10Sn wt. %)* for the active-substrate process and Incusil ABA (Ag-27.5Cu-12.1In-1.2Ti wt. %)** for the active-filler metal process. The solidus temperatures are 605°C for Braze 604 and 610°C for Incusil ABA and the liquidus temperature is 720°C for both filler metals.

The Mg-PSZ was ground using 320 grit and 400 grit SiC paper successively, and then polished on a Buehler Vibromet I Polisher with 1-micron diamond paste to produce a mirror finish. The NCI was ground using 320 grit SiC paper. Prior to brazing, filler metals were slightly abraded with a 600 grit SiC paper to disrupt oxide films on the surface and then ultrasonically cleaned with acetone followed by alcohol.

The Mg-PSZ subjected to the active substrate process was coated with Ti by one of three techniques; radio-frequency (RF) sputtering, physical vapor deposition using a tungsten basket and an electron-beam. RF-sputtering# was the principal technique and was used to deposit 0.3μm thick films on large PSZ specimens and four different thicknesses (0.3, 0.6, 1.0 and 2.0μm) on separate small specimens. As a result of problems encountered in early experiments (described in following section), some Ti-coating were deposited by physical vapor deposition using either a tungsten basket¶ or an electron-beam¶¶ to vaporize the Ti.

All of the brazing experiments were carried out in a graphite resistance heated vacuum hot press at The Ohio State University except one large contact area joint and the small contact area joints, which were brazed at Oak Ridge National Lab.

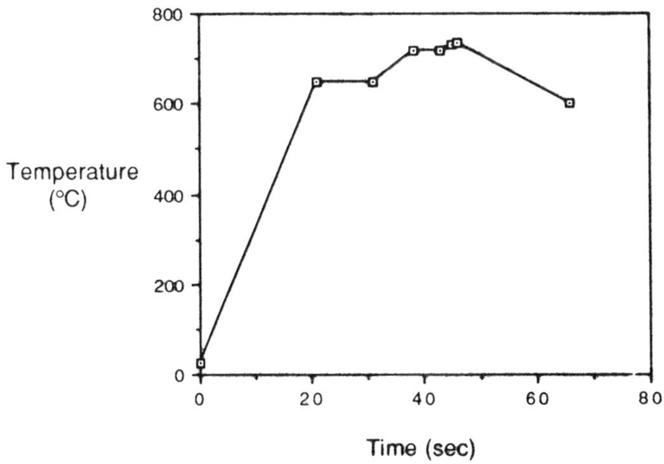

Figure 1. Brazing schedule for large contact-area joints.

* Lucas Milhaupt, Cudahy, WI
** GTE Wesgo, Belmont, CA
# Courtesy of Frank Honecy, NASA Lewis Research Center, Cleveland, OH
¶Edward Vacuum Evaporator
¶¶Courtesy of Michael Santella, Oak Ridge National Lab, Oak Ridge, TN

The typical brazing schedule for large contact area joining included heating up to 650°C at 30°C/min, holding for 10 min, heating to 720°C at 10°C/min, holding for 5 min, and heating to 730°C at 5°C/min where the power was shut off, see Figure 1. It was found experimentally that due to the thermal mass of the specimen and the radiant properties of the graphite element this procedure resulted in a maximum temperature very close to 735°C. The heating schedule used for small-contact-area joining included heating up to 650°C at 13°C/min and to 735°C at 3°C/min. There were no constant-temperature holding periods for these specimens. Temperature was measured using a Type K thermocouple inserted in a hole at the center of a NCI piece for the large specimens whereas for the small specimens a thermocouple was placed just above the specimens. Figure 2 schematically shows the specimen arrangement for the large-contact-area brazing.

The maximum brazing temperature was 735°C for the active-substrate process[9] and 750°C for the active-filler-metal process. The latter temperature was recommended by the brazing alloy manufacturer.*

Small alumina specimens (6.35 mm x 6.35 mm x 0.66 mm) cut from an alumina substrate§ were brazed to themselves by the active-filler-metal process using several heating schedules different from those of the PSZ brazing. Two Type K thermocouples were used to measure temperatures at two different locations; one adjacent to the specimens and the other in the middle of the heating zone, see Figure 3. Before brazing, the furnace temperature was first increased to 780°C. It was experimentally found that the ≈30°C overheat was necessary due to the specimen's thermal mass.

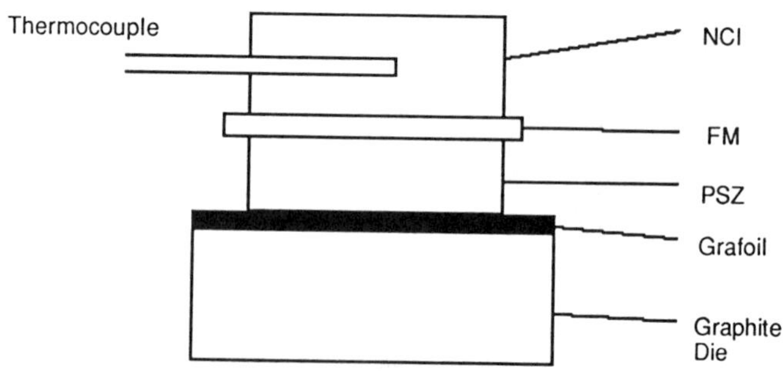

Figure 2. Specimen arrangement for brazing large contact-area joints.

---

* GTE Wesgo, Belmont, CA

§Coors, Golden, CO

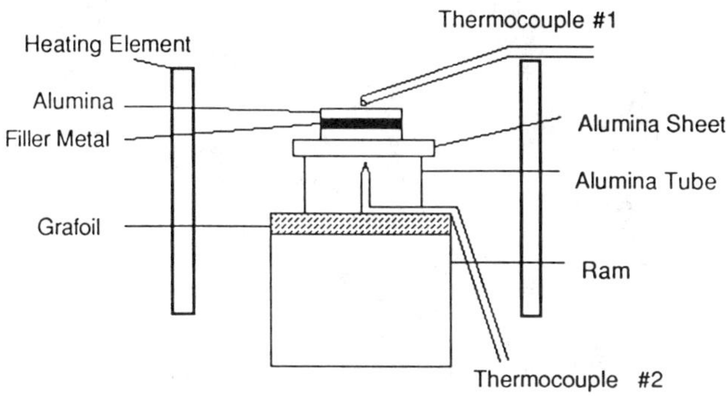

Figure 3. Specimen arrangement for brazing alumina joints.

The specimen was then raised into the heated zone where the specimen temperature reached 600°C, held for 5 min, raised to the 700°C zone of the furnace, held for 10 min, and further raised to the 750°C zone and held for a selected period of time. Hold times were 1, 5, 10, 30, 60 and 120 min. The specimen was then lowered out of the heating zone and the furnace was shut off.

Typical vacuums for all of the brazing experiments were in the range of $2\text{-}4 \times 10^{-5}$ torr ($2.7\text{-}5.3 \times 10^{-3}$ Pa).

Results

Large contact-area joints

The brazing of large PSZ to NCI joints using the active-substrate process was consistently unsuccessful. To examine the braze for the presence of voids , ultrasonic non-destructive evaluation using the C-scan method was performed at Battelle. The results of the C-scan examination revealed continuous voids, several mm wide, in the joint area, see Figure 4. The white areas indicate the brazed region and the dark areas denote voids in the brazed joint. Three bend bars were cut from the brazed area of this specimen and tested at Battelle. In four-point bend tests, they failed at 80.9, 88.9 and 81.4 MPa.

Figure 4. Ultrasonic C-scan image for a large PSZ to NCI joint. (Brazed at OSU)

In an effort to eliminate the voids, a sequence of experiments was conducted using a variety of methods to deposit the Ti-layer. In addition, two different brazing furnaces were used. It was first hypothesized, based on Winters and Kay[10], that Ar might be entrapped in the Ti-layer during sputtering and diffused out to form the voids during brazing. To test this, an RF-sputter deposited specimen was preannealed in vacuum to 735°C prior to brazing through the normal brazing cycle to allow any trapped Ar to escape. The C-scan image of a brazed joint fabricated using the PSZ with the preannealed Ti-layer was indistinguishable from that shown in Figure 4, strongly suggesting that Ar was not the source of the voids. This was further supported with C-scans of active-substrate joints fabricated using the PSZ with vapor deposited Ti layers that should not have any entrapped gases. C-scans from three such joints were very similar to Figure 4.

A further trial was attempted at Oak Ridge National Laboratory (ORNL) to determine if the vacuum hot press at OSU was responsible for the void generation. A set of RF-sputter coated PSZ and NCI specimens was brazed using standard ORNL procedures.* Although the resultant void distribution was qualitatively different than in the experiments conducted at OSU, see Figure 5, the result was similar in that less than 50% of the joint area was brazed in this experiment.

* Courtesy of Michael Santella, Oak Ridge National Laboratory, Oak Ridge, TN

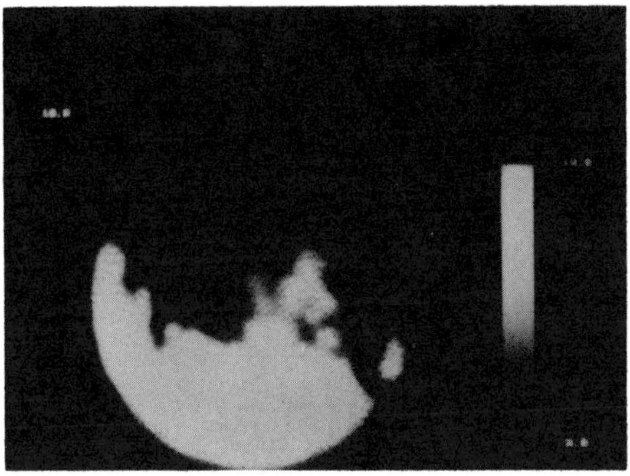

Figure 5. Ultrasonic C-scan image for a large PSZ to NCI joint. (Brazed at ORNL)

In order to observe the flow behavior of the filler metal and to determine when the voids are formed, a brazing experiment was attempted using the standard adherends, but with three foils, 38 mm in diameter, stacked and placed concentric to the 50-mm-diameter circular face of the PSZ. In this experimental configuration, the filler metal was thick and only covered the central 50% of the PSZ-NCI contact area initially. It was anticipated that the filler metal would therefore flow outward rapidly upon melting, which was expected to help remove any impurities and to fill the gap. The brazed joints revealed that the outward flow of the molten filler metal did occur, indicating complete melting; however, it did not fill the gap between the PSZ and the NCI. The filler metal ran out in narrow channels and collected at the corners of the NCI block that were not covered by the PSZ disc. These results indicated that the Ti reacted to form non-wettable surface before the filler metal melted.

In contrast to the results of the active substrate process, the brazing of PSZ to NCI using the active filler metal process was routinely successful producing strong joints with no voids. Four bend bars were cut from these specimens and tested in four-point bending exhibited strengths of 211, 214, 320 and 327 MPa.*

* Courtesy of Bashkar Majumdar, Battelle, Columbus, OH

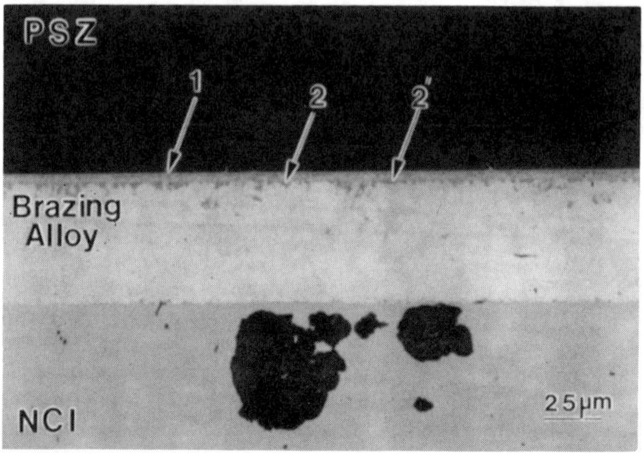

Figure 6. Optical photomicrograph of a PSZ to NCI joint brazed by the active-substrate process with 2μm Ti.

Table 1 EPMA Analysis for the area near the PSZ surface

	Elements (at.%)*				
area	Ti	Sn	Cu	Ag	O
1	39.8	1.9	1.8	-	54.7
2	13.2	12.8	49.7	1.1	23.1
2'	26.3	16.1	23.6	6.9	27.1

*Trace elements are ignored.

Small contact-area joints

The four Mg-PSZ joints with different thicknesses of RF-sputtered Ti (0.3, 0.6, 1.0 and 2.0 μm) were brazed successfully using the active substrate process. Figure 6 is an optical photomicrograph of the braze between PSZ and NCI with 2.0 μm Ti. Table 1 shows a summary of data obtained using WDS quantitative analysis on the different phases formed near the PSZ to filler metal interface.

An interesting feature is the darkened layer formed at the PSZ surface after brazing, as seen in Figure 7. These layers are believed to result from reduction of the PSZ by the Ti-layer. However, quantitative analysis using EPMA was unable to detect these small changes in oxygen concentration. The thickness of this layer, measured by an optical microscope, was linearly related to the original Ti thickness, see Figure 8.

In a separate set of experiments, alumina was successfully brazed to alumina using the active-filler-metal process. Specimens were held at the brazing temperature, 750°C, for various times (1, 5, 10, 30, 60 and 120 min) to determine whether reactions between the Ti and alumina affected joining. In all specimens, strong joints were formed and the interlayer was free of porosity.

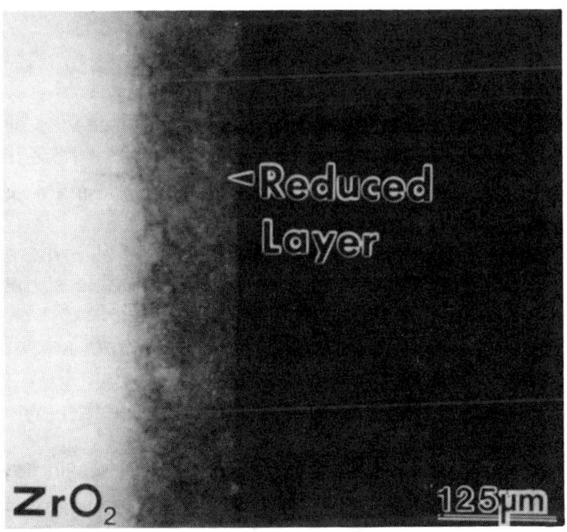

Figure 7. Optical photomicrograph showing a dark (reduced) layer near the PSZ surface.

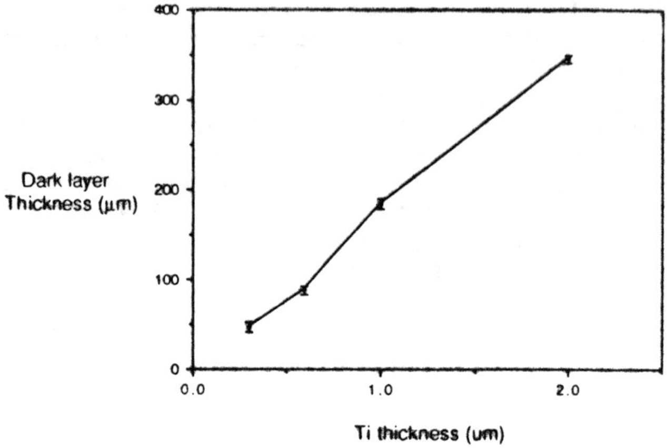

Figure 8. Dark layer thickness vs. Ti layer thickness.

Discussion

The void formation in the large active substrate joints was similar to Patten et al.[11] who also found porosity in the filler-metal layer of PSZ to gray cast iron joints brazed using Incusil 15 ABA.* Their efforts to eliminate porosity by using different furnaces also were unsuccessful.

The results of the experiment using three stacked filler-metal foils indicated that the formation of the voids was due to a non-wettable surface on the PSZ. This non-wettable surface is presumed to be the result of oxidation of the Ti layer. Naidich[1] and Nicholas[12] reported that the contact angle of liquid Cu on TiO_2 and Ti_2O_3 were 150° and 115°, respectively, while that on TiO was 70°. These data indicate that the Ti/O ratio of the Ti-rich reaction layer must be equal to or greater than 1 for wetting.

There are three possible sources for oxygen in the brazing system: the filler metals, the PSZ and the atmosphere. These possibilities are discussed below.

In the active-substrate process Ti is present with high specific surface area and is in direct contact with both the atmosphere and the ceramic, while in the active-filler-metal process Ti is present as a Ti-Cu alloy[13] in Incusil ABA and is completely surrounded by the filler metal. Although it is possible that the filler metal is an oxygen source, it is unlikely because the solubility limits of oxygen in

*Ag-23.5Cu-14.5In-1.25Ti wt. %.

Ag[14] and Cu[15] are low and, if the Ag-Cu phase were a significant source of oxygen, the Ti phase distributed in the Incusil ABA would be oxidized. This was not observed.

The darkened layers formed at the PSZ surface, see Figure 7, suggest that the PSZ is being reduced, i.e. the ceramic is an active source of oxygen. A rigorous calculation using the relevant material properties supports this idea.

Zirconia tolerates nonstoichiometry by the following equation

$$ZrO_2 = ZrO_{2-x} + (x/2)O_2. \tag{1}$$

To produce 1 mol of the wettable TiO, 1/x mol of ZrO_2 must be reduced and to produce 1 mol of the nonwettable Ti_2O_3, 1.5/x mol of ZrO_2 must be reduced. Similarly for Ti to be completely oxidized to TiO_2, 2/x mole of ZrO_2 should be reduced. The following calculation is based on the assumption that the nonwettable surface is TiO_2.

The number of moles of reduced ZrO_2, n_z, is given by

$$n_z = \frac{V_z\, d_z}{M_z} = \frac{A_z\, t_z\, d_z}{M_z} \tag{2}$$

where V_z, A_z, and t_z are the volume, the area, and the thickness of reduced ZrO_2 and d_z and M_z are the density and the molecular weight of ZrO_2. The molar ratio of ZrO_2 to Ti is then

$$\frac{n_z}{n_t} = \frac{A_z\, t_z\, d_z\, M_t}{A_t\, t_t\, d_t\, M_z} \tag{3}$$

where n_t is the number of moles of Ti. Because the areas of ZrO_2 and Ti are the same, i.e. the coating completely covers the surface, $A_z = A_t$ and Eqn. (3) is rewritten as,

$$\frac{2}{x} = \frac{t_z\, d_z\, M_t}{t_t\, d_t\, M_z} \tag{4}$$

Solving for t_z gives

$$t_z = \frac{2\, t_t\, M_z\, d_t}{x\, M_t\, d_z} \tag{5}$$

If the thickness of the Ti coating (t_t) is 1 μm; x is assumed to be 10^{-2};[16] and values are substituted for M_z, M_t, d_z and d_t; the thickness of the reduced layer, t_z, is calculated to be 400 μm. An equivalent calculation to form Ti_2O_3 gives a 300μm-thick reduced layer. These numbers are not far from the measured thickness of the darkened layer, 185μm, indicating that the ceramic is a significant contributor to Ti oxidation. It should be noted that the value of the deviation from stoichiometry, x($\tilde{}$ 10^{-2}), is estimated as the upper limit by the equilibrium oxygen partial pressure of TiO_2 at the brazing temperature and the data from Kofstad.[16]

The atmosphere also may supply oxygen in addition to that contributed by the ceramic in the active-substrate process as long as there is a finite gap between the Ti-coating and the filler metal. In order for the atmosphere to be an effective source for Ti oxidation, the oxygen partial pressure inside the furnace must be higher than the equilibrium partial pressure of oxygen for the TiO_2 formation. A theoretical prediction of the oxygen pressure in the furnace can be made by studying the thermodynamics of the C-O system provided by Estell and Flengas.[17] The theoretical oxygen pressure is given by

$$\log(P_{O_2}) = 2(P_{CO_2}) - 2\log(P_{CO}) - 29502.8/T + 9.0694 \qquad (6)$$

where, $$2(P_{CO}) = -k + (k^2 + 4kP_t)^{1/2} \qquad (7)$$

$$2(P_{CO_2}) = k + 2P_t - (k^2 + 4kP_t)^{1/2} \qquad (8)$$

$$\log k = -8916.4/T + 9.1131 \qquad (9)$$

and P_t = total oxygen pressure in the furnace.

The total oxygen pressure, P_t (2.8×10^{-9} atm), is given by assuming it to be 21% of the measured pressure of the furnace (10^{-5} torr = 1.3×10^{-8} atm). Substituting values for k at the brazing temperature (735°C) and P_t gives the equilibrium Pco and P_{CO_2} and, accordingly, P_{O_2}. The oxygen pressure obtained from these equations is $\approx 2 \times 10^{-28}$ atm. Since this value is more than ten orders of magnitude greater than the calculated equilibrium oxygen partial pressure for TiO_2 formation[18] (10^{-40} atm), it is obvious that the exposed Ti will be oxidized by the atmosphere.

Once the filler metal melts, the gap is filled and atmospheric oxygen must diffuse through the filler metal to oxidize the Ti. This is the case throughout the active-filler-metal brazing process as the Ti is always surrounded by the filler metal. A calculation based on the permeability, i.e. the product of the diffusivity and the solubility of oxygen in Ag (10^{-3} % by mass)[19] at the relevant oxygen pressure ($\approx 10^{-28}$ atm, see Eqn. 6) gives a very low value of the oxygen flux ($\approx 10^{-24}$ mols/cm^2 sec) through the filler metal in either solid or liquid states, suggesting that it will serve as an effective diffusion barrier. This calculation is supported by the results of the alumina brazing experiments.

Brazing of alumina specimens was performed to check the effectiveness of the filler metal as a diffusion barrier. No evidence of dewetting was found even for specimens held for long times up to 120 min at the brazing temperature. In this system, the alumina cannot supply oxygen because it is a line compound that does not tolerate non-stoichiometry and the oxygen diffusion is too low.[20] Therefore, the only source of oxygen is the atmosphere. However, because active-filler-metal was used in this system, oxidation required diffusion of oxygen through the filler metal, whether solid or liquid.

Conclusions

Voids in the PSZ to NCI joints brazed by the active-substrate process are due to the formation of a non-wettable surface on the PSZ, which is presumed to be the result of Ti oxidation.

There are two potential sources of oxygen. The PSZ appears to be significant as long as the Ti is in direct contact with it. In the active-filler-metal process, the PSZ is a significant oxygen source only after melting of the brazing alloy. In the experiments reported here that used the active-substrate brazing process, the brazing atmosphere also may be an important oxygen source before the filler metal melts but not after.

The different results obtained in brazing the PSZ compared to brazing the alumina are due to the active filler metal playing a role as a diffusion barrier and the different transport properties between the PSZ and the alumina.

Acknowledgements
The authors at OSU are grateful to Battelle for technical support and to EWI and Oak Ridge National Laboratory for financial support.

References

1. Y. V. Naidich et al., "Adhesion, Wetting, and Formation of Intermediate Phase in Systems Composed of a Titanium–Containing Melts and an Oxide," Sov. Powder Metall. Metal Ceram., 12 (11) (1973), 895–899.

2. C. W. Fox and G. M. Slaughter, "Brazing of Ceramics," Welding J., 43 (7) (1964), 591–597.

3. H. Mizuhara and E. Huebel, "Joining Ceramic to Metal with Ductile Active Filler Metal," Welding J., 65 (10) (1986), 43–51.

4. A. J. Moorehead and H. Keating, "Direct Brazing of Ceramics for Advanced Heavy Duty Diesels," Welding J., 65 (10) (1986), 17–31.

5. P. Martineau et al., "SiC filament/titanium matrix composites regarded as model composites," J. Materials Sci., 10 (1984), 2749–2770.

6. Takashi Tanaka, Hiroshi Morimoto, and Hiroyuki Homma, "Joining of Ceramics to Metals," Nippon Steel Technical Report 1988, No 37:31–38.

7. P. M. Scott, M. Nicholas, and B. Dewar, "The wetting and bonding of diamonds by copper–base binary alloys," J. Materials Sci., 10 (1975), 1833–1840.

8. A. J. Moorehead, H. M. Henson, and T. J. Henson, "The Role of Interfacial Reactions on the Mechanical Properties of Ceramic Brazements," Ceramic Microstructures '86, ed. A. G. Evans (Plenum, 1988), 949-958.

9. J. P. Hammond, S. A. David, and J. J. Woodhouse, "Indirect Brazing of Structural Ceramics for Uncooled Diesels", Proceedings of 22nd Automotive Engineers, 1985, 523-529.

10. H. F. Winters and E. Kay, "Gas Incorporation into Sputtered Films," J. Appl. Phy., 38 (10) (1967), 3928-3934.

11. Donald O. Patten et al., "Mechanical Behavior of Ceramic-Metal Braze Joints," Ceram. Eng. Sci. Proc., 10 (11-12) (1989), 1866-1878.

12. M. G. Nicholas, "Active Metal Brazing," British Ceram. Trans. & J., 85 (4) (1986), 144-146.

13. Eunsung Park, "Active Metal Brazing of Ceramics," (MS thesis, The Ohio State University, 1990), 83.

14. Hansen, Constitution of Binary Alloy (New York, NY: McGraw-Hill, Inc. 1958), 37.

15. J. P. Neumann, T. Zhong, and Y. A. Chang, "The Cu-O (Copper-Oxygen) System," Bulletin of Alloy Phase Diagram, 5 (2) (1984), 138-140.

16. Per Kofstad, Nonstoichiometry, Diffusion, and Electrical Conductivity in Binary Metal Oxides (New York, NY: John Wiley & Sons, Inc., 1972), 154

17. T. H. Estell and S. N. Flengas, "The Determination of Oxygen in Gas Mixtures by Electromotive Force Measurements Using Solid Oxide Electrolytes," Metallurgical Transactions, 3 (1) (1972), 27-36.

18. JANAF Thermochemical Tables, U.S. National Bureau of Standards, 1970

19. E. A. Brandes, ed., Smithell's Metal Reference Book (London: Butterworths, Inc. 1983), 12-19.

20. W. D. Kingery, H. K. Bowen, and D. R. Uhlmann, Introduction to Ceramics (New York, NY: John Wiley & Sons, Inc., 1960), 240.

SILICATE BRAZING OF ALUMINA CERAMICS USING CALCIUM ALUMINOSILICATE INTERLAYERS

Jeffrey T. Knapp and James D. Cawley

Department of Materials Science and Engineering
The Ohio State University
2041 College Road
Columbus OH 43210-1178

Abstract

Silicate brazing of alumina ceramics was investigated. A brazing system based on a glass with a chemical composition equivalent to the mineral anorthite ($CaAl_2Si_2O_8$) was developed and tested. This system proved successful and average bend strengths of ≈145 MPa were achieved with vitreous joints processed at 1600°C for 0.5 hour. Room temperature strengths were retained to 500°C. Crystallization of the interlayer was achieved using a controlled post-brazing heat treatment. Contrary to expectation, the crystallization degraded the performance of the system.

This work was supported through the Cooperative Research Program of the Edison Welding Institute, 1100 Kinnear Road, Columbus OH 43212.

Metal - Ceramic Joining
Edited by P. Kumar and V. A. Greenhut
The Minerals, Metals & Materials Society, 1991

INTRODUCTION

A recognized limitation of ceramic components is that the strength, density, and uniformity of these parts tend to decrease with size [1]. This favors the joining of smaller ceramic parts to produce final components. Two other possible requirements of the system [2] also may lead to the need to join two or more ceramic components. Oftentimes, only simple shapes can be made by the primary fabrication process and secondary processes such as joining or machining are necessary to obtain a desired complex shape. Another possibility is that an application requires two or more dissimilar materials joined together to create a single component.

One important method for joining ceramic materials is brazing [2], usually identified on the basis of the interlayer material used, either metal or ceramic [3]. In this research two silicate brazing systems for joining high-purity alumina were evaluated.

BACKGROUND

Metal brazing of ceramics to ceramics and to metals is a widely known technique [1,2, 4, 5], and an example of an application that has been developed since 1950 is the joining of alumina to metal in vacuum tubes [6]. Other applications in which alumina was being joined have been reviewed [7]. The primary constraint [8] on most commercial brazing filler metals employed in metallurgical applications is that they will not wet ceramic oxides. As a result certain processes have been developed that allow the metal braze to wet the ceramic. These processes have been categorized [2] as noble-metal brazing, active-metal brazing, and refractory-metal brazing.

An example of the joint strengths that can be achieved via metal brazing is found in work done by Moorhead [8]. In this work a braze with a composition of 45 Cu-26 Ag-29 Ti, at. %, was used to join Coors AD-998 alumina. The alumina alone was shown to have a four-point bend strength of 207 MPa, while the joined alumina broke under an average stress of 222 MPa at room temperature. At 400°C the flexural strength of these joints was found to be 215 MPa and at 600°C found to be 213 MPa.

Brazing using a silicate interlayer is similar to metal brazing but differs in that achieving wetting does not generally pose a problem [2,3]. Ceramic brazes often offer better environmental compatibility than metals, but, in general, still not equal to that of the base material [2]. Ceramic brazes are usually less tolerant of thermal expansion mismatch but may better match the thermal expansion coefficient of the ceramic adherends [2,9].

The effect of joint thickness and thermal expansion mismatch on the mechanical properties of joints made with silicate brazes was studied by Kirchner and others [10, 11]. This study consisted of finite element analysis to estimate the stresses involved [10] and mechanical testing of alumina-to-alumina butt joints with interlayers of various thicknesses and with varying degrees of thermal expansion mismatch [11]. The stress analysis [10] for brazes with thermal expansion coefficients lower than that of the adherend, which is relevant to the current study, revealed that axial stresses are tensile and may result in lower fracture stresses.

The magnitude of such stresses tends to be greater near the midplane of the interlayer and near the perimeter of the joint. These tensile stresses also tend to increase with increasing thickness.

Alumina is an important technical ceramic and consequently joining of alumina has been the subject of a number of experimental studies. Hauth [3, 12] has reported that ceramic sealant bonding, using both vitreous and crystalline materials, may be used for joining technical ceramics into large, complex shapes. Several glass sealants with joining temperatures in the range of 450°C to 1500°C have been developed that are suitable for joining alumina to alumina or to sapphire. The high-temperature sealants [12] reported were based on kaolin and on potassium feldspar, while the low-temperature glasses [12] were all borate compositions.

The Al_2O_3-MnO-SiO_2 [13] and CaO-MgO-Al_2O_3-SiO_2 [14] systems were researched by Klomp and Botden. Joints from both systems were evaluated in terms of their room-temperature strength and their softening temperature. The latter indicated that the seals could be routinely subjected to temperatures as high as 750°C. The Al_2O_3-MnO-SiO_2 [13] system was used to produce alumina seals with average three-point bend strengths as high as 178 MPa and Kovar-alumina seals with strengths as high as 184 MPa. It was also noted in this study that certain seal compositions almost completely crystallized, resulting in seals with considerably lower strengths than those which remained vitreous.

Alumina-alumina joints were also formed using mixtures from the CaO-MgO-Al_2O_3-SiO_2 [14] system. For four different compositions, the melting point of the mixture, the softening temperature of the seal, and the average three-point bend strength of the joints are listed in Table I. The increase in softening temperature over the original melting temperature was a result of dissolution of alumina in the melt during sealing. It was also believed that the growth of alumina whiskers into the braze improved bend strengths [14].

Table I. Characteristics of compositions from calcia-magnesia-alumina-silica system used as brazes for joining alumina.

Batch Mixture No.	Composition (wt %) CaO	Al_2O_3	MgO	SiO_2	Batch Melting Temp.(°C)	Joint Softening Temp.(°C)	Failure Stress (MPa)
1	42	18	0	40	1300	1400	150
2	46	48	6	0	1350	1600	110
3	0	35	15	50	1500	1590	160
4	37.5	0	6.5	56	1400	1420	220

In another investigation [15], several brazes capable of being crystallized were used to join alumina. These joints exhibited fracture toughnesses (K_{IC}) in the range of 1.1 to 2.3 MPa•$m^{1/2}$. Joining of alumina at 1500°C with a 50/50 (by weight) mixture of talc and alumina produced the toughest joints of those tested. This talc/alumina mixture was chosen so as to form a composition that closely matched the grain-boundary phase of the alumina being joined. However, the joints produced with this mixture contained a large number of pores.

Clark and Reed [16] bonded alumina abrasives with glass-ceramics by sintering glasses of fine particle size in contact with alumina. By choosing the systems properly and using a chemical processing step to etch the surfaces of the glass powders to reduce the number of nucleation sites, the glass would successfully undergo viscous sintering with the alumina during fabrication and could be crystallized after forming was complete. They found the etching step useful because sintering and crystallization are competing processes if they are allowed to occur simultaneously. The most successful composition proved to be one whose major crystalline phase was parawollastonite (ß-$CaSiO_3$). This composition yielded glass-bonded alumina with a diametral compression strength of 66 MPa. In contrast to the deleterious effects of crystallization noted by Klomp and Botden, in this work fractography revealed that the wollastonite crystals which were produced by devitrifying the glass increased the resistance of the material to crack growth.

There have been three different approaches used to apply the interlayer material. The most common method, and the one selected for this research, is to apply the interlayer in a powdered form between the adherends [11, 14, 15]. However, use of a powdered glass may result in the formation of pores if initial coverage is incomplete or as a result of gas trapped in the original melt.

Two techniques have been used to avoid these problems. One method [17] was to glaze one of the adherends with a slurry of powdered glass and to grind the fired glaze layer to the appropriate thickness before joining the assembly with the second adherend. The second was to use a thin plate of glass of known thickness between the two pieces being joined [11].

The braze material needs to be easily vitrified into a homogeneous glass that is stable over a wide temperature range and one which melts at a high temperature in order to produce joints that are both strong and corrosion resistant. At the joining temperature the liquid formed should wet the alumina and penetrate into the grain boundaries. Wetting of the alumina allows for bond formation, and penetration of the glass reduces the thickness of the interlayer. Studies have been performed [11, 17] that show that reduction of the interlayer thickness results in stronger joints. The final step of a post-joining crystallization anneal should increase the refractoriness of the joint and the resistance to static fatigue if the glass behaves like certain devitrifying solder glasses [4]. Figure 1 schematically illustrates the microstructural changes which occur during brazing with such a silicate: a) a powdered glass layer is placed between the adherends; b) at the brazing temperature the glass is vitrified and penetrates the grain boundaries; c) in a separate (optional) annealing step the glass is devitrified.

Two systems were examined in this investigation: one based on talc and alumina which proved unsuccessful and one based on anorthite. The talc-based work is summarized elsewhere [18]. This paper is devoted to the characterization of an anorthite-based braze system.

Anorthite glass, An-glass, proved to meet the qualifications listed above. From the work of Dogan [19], a primary phase of anorthite was found to readily crystallize in the grain boundaries of a 96% alumina after an anneal of 1.0 hr at 1100°C. Other researchers [20-22] have established that glasses with compositions near that of anorthite can wet and penetrate the grain boundaries of polycrystalline alumina. Solution and precipitation reactions, predicted for this brazing system on the basis of the phase diagram [23], have been shown to have a

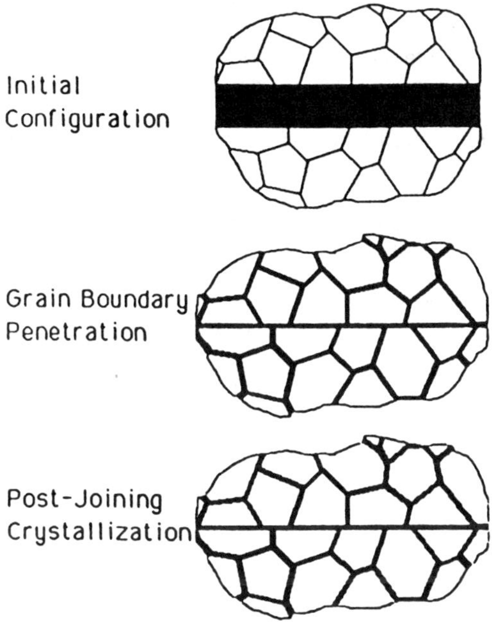

Figure 1. This figure is a schematic of the sequence of steps involved in silicate brazing: a) initially a powdered glass layer of constant thickness i placed between the adherends; b) upon heating the glass is melted, densifies, and penetrates the grain boundaries drawing the surfaces together; c) crystallization of the interlayer and grain boundary phase is accomplished through a post-joining controlled heat treatment.

direct effect on the capacity of a composition to wet and penetrate the grain boundaries of alumina [20, 21]. In addition, An-glass has a high melting point of 1553°C [23] and consists of oxides that have low vapor pressures and are stable in their oxygen content [24].

There is however an appreciable mismatch in coefficient of thermal expansion (CTE). The CTE for An-glass is ≈5.9 x 10^{-6} /°C [25] in contrast to 8.8 x 10^{-6} /°C [26] for polycrystalline alumina. Furthermore the thermal expansion coefficient of anorthite is highly anisotropic. Kozu and Ueda [25] reported the thermal expansion coefficient for a plagioclase consisting of 95 mole percent anorthite to be 5.9x10^{-6}/°C in the [100] direction, 2.4x10^{-6}/°C perpendicular to the (010), and 5.8x10^{-6}/°C perpendicular to the (001).

MATERIALS AND EXPERIMENTAL PROCEDURE

The raw materials used in making An-glass were kaolin, nominally $Al_2Si_2O_5(OH)_4$, and whiting, $CaCO_3$ (Columbus Clay Co., Columbus OH). A cellulose ether, Methocel™ (Dow Chemical Co., Midland MI), was used as a binder in the recommended ratio of 10g per kilogram of dry An-glass [27]. The alumina used as the adherend material was AD-998 (Coors Ceramic Co., Golden CO). The alumina used was in the form of rods ≈2 cm in diameter. This alumina was independently identified as having a four-point flexural strength of 207 MPa [8], while the manufacturer reported its room temperature strength to be 331 MPa. Presumably this difference arises from surface preparation or testing method. In this study, the base strength of the adherend was assumed to be the lower of these values. Samples, ≈2.5x2.5x0.06 cm, cut from Coors 995 alumina substrates were used for glazing experiments.

The An-glass batch was made by mixing equal molar amounts of kaolin (258.1 g/mol) and whiting (100.09 g/mol) using a mortar and pestle. This mixture was melted for 0.5 hr at 1600°C, typically in a platinum crucible. In a limited number of cases melting was accomplished in an alumina vessel.

The glass was thermally shocked and broken to remove it from the crucible. Most of the glass was ground to a fine powder for use as the interlayer material although some bulk specimens were prepared for crystallization studies and mechanical testing.

The powdered glass and the Methocel™ were mixed as dry ingredients. This mixture was then made into a paste by adding approximately 1.7 ml of distilled water to every 2 g of mix. This paste was easily spread over the alumina surface. In joining alumina to alumina one adherend would first be glazed with the wet mixture and the second one would be placed on top of the wet glaze. This green state join would typically be dried for at least 45 minutes before being fired.

An initial set of experiments determined that ≈62 mg/cm^2 is the optimal amount of dry mixture that should be used in joining. This amount of powder was equivalent to a coverage of 62 mg/cm^2 of braze mixture. This amount of powder yielded fired joints that were consistently joined over a significant fraction of the adherend diameter. Use of powder amounts greater than the optimum such as 78 or 93 mg/cm^2 resulted in partial loss of the interlayer material during wet-state processing. This loss would occur because part of the paste would be forced out of the joint when the upper adherend was placed on the wet interlayer. Powder amounts of 50 and 31 mg/cm^2 were also used and often yielded incomplete interlayers.

The joined and the glazed samples were fired in air in a furnace programmed to ramp to 1600°C (≈50°C above the melting point of anorthite) at a rate of 10°C/min. Samples were held at 1600°C for either 0.5 hr or 2.0 hr. The furnace was typically programmed to cool at a rate of 16°C/min and, therefore, to reach room temperature in 100 min. However, it was observed that the furnace lagged the program and the actual cooling rate was between 11 and 16°C/min. Devitrification was not visually observed in glass melted in a platinum crucible and cooled in this manner.

Crystallization of selected specimens was carried out in a separate post-brazing anneal of 0.5 hr at 1100°C. This annealing schedule is similar to that used by Dogan to crystallize anorthite in the grain boundaries in a type of 96% alumina [19]. The heating and cooling rates were the same as those used during joining.

Prior to joining, adherends were ground to ensure that the ends were flat and parallel using a precision handgrinder (South Bay Technology, Temple City CA). All mechanical testing was completed using a Model 4201 Instron testing machine. Samples that were broken at 500°C were heated to temperature using an induction-coil furnace system that was assembled in-house. The four-point bend strength specimens had a butt-joint configuration and were machined (Bomas Machine Specialties, Somerville MA) to meet or exceed military standards [28]. The final grit size used on the tensile surface was 320 grit with the exception of two sets of bars each of was ground to 900 grit as recommended by Johnson and Rowcliffe [17]. The same loading conditions were used in all experiments. At both temperatures the spans used were 13.0 mm and 32.5 mm, and the cross head displacement rate was 0.5 mm/min.

Bend strengths are reported for three sets of specimens. The interlayer application was 62 mg/cm^2 for all three sets. During joining Set-A was held at 1600°C for 2 hrs while Set-B and Set-C were held for 0.5 hr. Set-C was subjected to a devitrification anneal at 1100°C for 0.5 hr.

All surfaces that were microstructurally examined were polished to a mirror finish with 3-µm diamond paste and further polished to 1 µm. Most samples were given a short finishing step at 1/4 µm. A subset of samples was first ground using 70 µm diamond and subjected to additional intermediate steps at 15 µm and 6 µm. Optical microscopy was performed on either a Zeiss optical microscope or a Nikon Optiphot. Most of the electron microscopy and all of the quantitative analysis was conducted using an electron microprobe (Cameca Camevax, Paris France). The microprobe was also used to obtain elemental X-ray maps of certain samples. A JEOL scanning electron microscope was used to supplement the EMPA. X-ray diffraction patterns of certain samples were obtained using a Philips X-ray diffractometer.

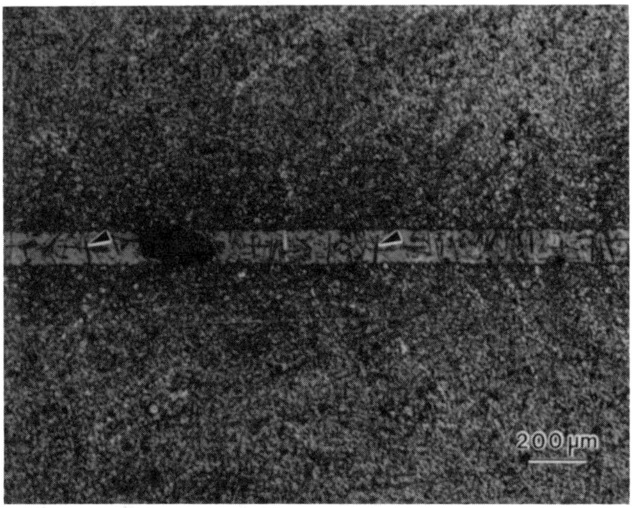

Figure 2. Optical micrograph (50 X, Nomarski imaging) of an alumina/ An-glass/ alumina joint that was formed using a 0.5 hr anneal at 1600°C. Arrows point to alumina crystals that have grown into the interlayer.

RESULTS

Strong joints were routinely produced using the An-glass braze consistent with low magnification optical micrographs which indicate a low porosity level of 5 to 10 percent within the braze layer of such joints. Figure 2 is an example of a typical joint. The overall porosity level is low, however there is an occasional pore that is the width of the joint and more than 100 μm in length.

A whiskery phase within the braze matrix is also evident in Figure 2, and at a slightly higher magnification in Figure 3. The latter shows that these acicular crystals are extensions from grains on the adherend surfaces and are 50 to 100 μm long and ≈2 μm wide. Some of these whiskers traversed cracks within the braze matrix. Electron probe microanalysis (EPMA) was used to determine that the whiskers are alumina. Roughly 42% of the whiskers were aligned nearly perpendicular (80-90°) to the adherend surface and 73% of them were within 60-90°. By dividing the area occupied by the crystals by the total braze area, an estimate of the volume fraction of whiskers was obtained and converted to weight percent. The alumina crystals were determined to constitute 5.7 weight % of the braze.

The microstructure of devitrified joints was very similar to that of the vitreous joints, although an increase in the density of cracks parallel to the adherend surfaces was noted.

Figure 3. Optical micrograph (200 X, Nomarski imaging) of an alumina/ An-glass/ alumina joint made by using 62 mg/cm^2 of initial braze mixture and firing the assembly at 1600°C for 0.5 hr. This micrograph was taken near the edge of the specimen where the interlayer thickness was a maximum. It is clear that the acicular alumina crystal are the result of epitaxial growth on a subset of the grains exposed on the adherend surface.

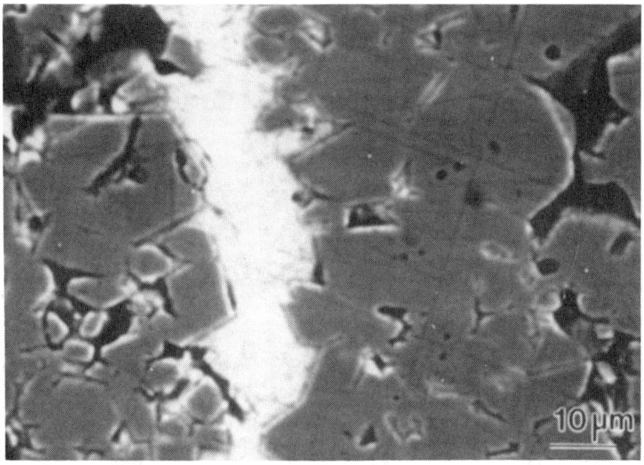

Figure 4. Backscattered electron image (1000 X) from the center of an alumina joint formed using an initial amount of braze mixture of 62 mg/cm^2 and using a joining step of 0.5 hr at 1600°C. Arrows point to glass that has penetrated into the grain boundaries.

This difference between the calculated and actual joint thicknesses requires transport of glass away from the joint. Penetration of the glass into the grain boundaries can be seen in Figure 4 but is better illustrated with elemental mapping. Figures 5a and 5b are elemental X-ray maps for Si and Ca in the region shown in Figure 4. These two maps, which superimpose on one another, disclose the penetration of the glass into the grain boundaries of the alumina. These maps show that a significant amount of glass had penetrated at least 60 μm into the adherends.

This was confirmed by a spot compositional analysis for Si and Ca at stepped distances from the interlayer. These results, see Figure 6, indicate that the glass concentration is appreciable at distances up to several thousand microns away from the original interface. Further analysis using a small spot-size (1 μm^2) beam was performed to measure the molar ratio of CaO to SiO_2 within the glassy phase as a function of distance from one of the An-glass/alumina interfaces. A plot of these data is shown in Figure 7. The five points closest to the joint interface represent the composition at five separate locations within the braze layer and indicate the calcia to silica ratio is constant across the braze. The other points plot the ratio in the glassy phase at various depths of penetration in the adherend alumina. Although there is significant scatter (presumably due to poor counting statistics associated with the small samples of glass), this figure shows that the ratio of these two components was not highly dependent upon the distance the glass had penetrated.

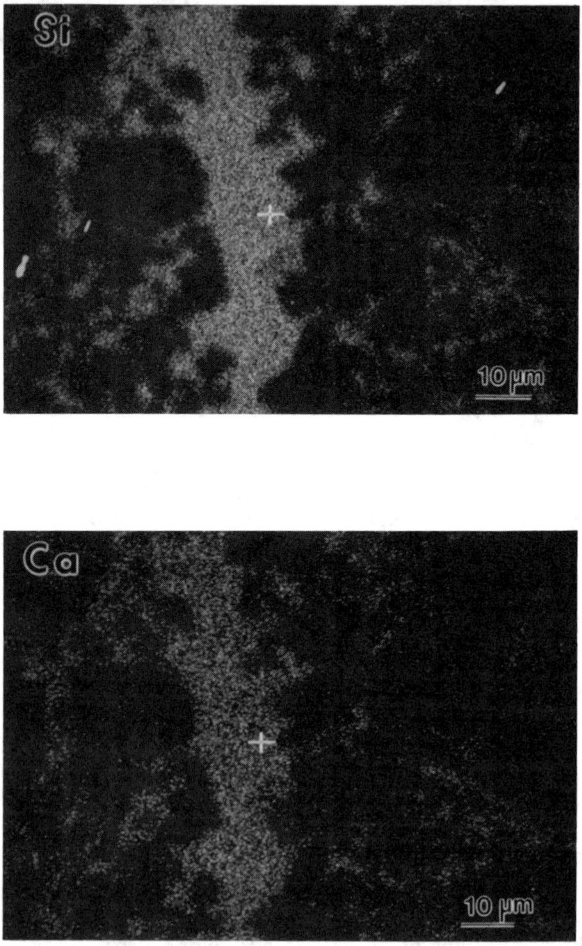

Figure 5. Elemental mapping (WDS) for: a) Si and b) Ca corresponding to the microstructure presented in Figure 4. Penetration of the interlayer glass into the grain boundaries is evident.

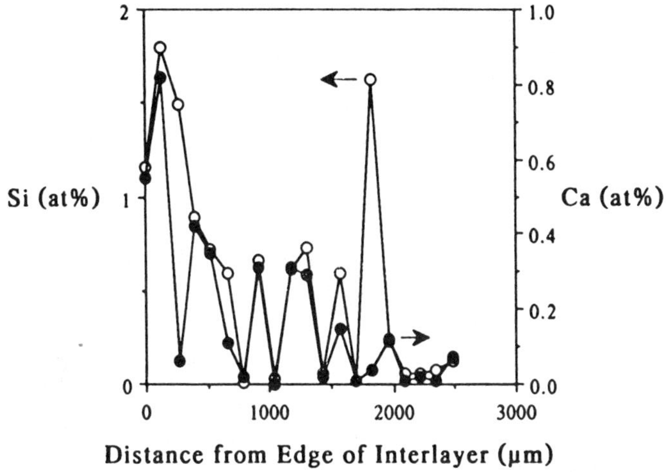

Figure 6. Concentration of calcium and silicon as a function of distance into the alumina in a sample annealed at 1600°C for 0.5 hr.

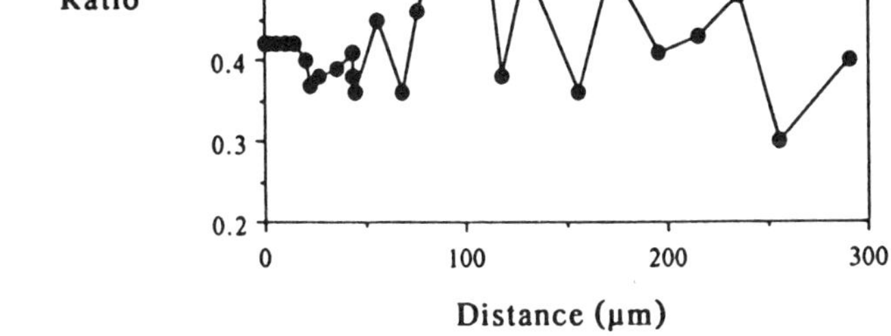

Figure 7. CaO / SiO_2 ratio as a function of position in a sample annealed at 1600°C for 0.5 hr. Zero on the distance axis represents an interlayer/ alumina interface and the first five points correspond to data taken from the interlayer.

Samples produced using 62 mg/cm^2 but with a 2.0 hr anneal at 1600°C were found to have continuous alumina grains at the joint interface, but also a relatively high degree of porosity, see Figure 8. EPMA results indicated that there was essentially no residual glass in the joint. This is believed to occur because the roughness of the original adherend surface effectively defines the diameter of the glass pockets formed when the surfaces are drawn together and establish rigid contact. If the diameters of these are sufficiently large, a capillary pressure will exist and the glass pockets will be drained by further glass flow into the alumina grain boundaries.

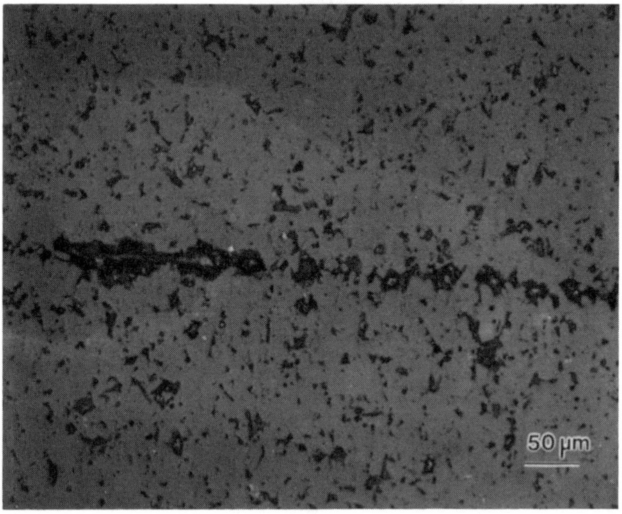

Figure 8. Optical micrograph (200X, polarized light) of an alumina/ An-glass/ alumina joint held at 1600°C for 2.0 hr. There is direct joining of alumina grains, but substantial porosity exists in the interlayer.

The modulus of rupture determined from four point bend tests proved to be dependent on the processing history. Six bars each from Sets -A, -B and -C were tested at room temperature and the results are presented in Table II. The average strengths were 91 MPa for Set-A, 148 MPa for Set-B, and 135 MPa for Set-C. These data were characterized by relatively large standard deviations, 40 -74 MPa.

Fractography of the specimens indicates that occasional large pores served as fracture origins and that the strongest joints (<100 MPa) were characterized by an oscillatory crack path, i.e. the crack propagated partially through the interlayer and partially through the alumina adherend. Figure 9 displays a region in which the fracture path was through the joint. A large area (about 50 µm across) can be seen in which alumina grains are apparent indicating coverage in the green state was incomplete. The nature of the crack trajectory is included in the Table.

Table II. Mechanical testing results using four-point bend method at room temperature on An-glass joints. The notation "j" indicates that the failure path was through the joint and "j/a" indicates that path at veered from the interlayer into the alumina.

Set-A (T_{join}=1600°C, t_{join}=2.0 hr.)

Sample Number	MOR (MPa)	Path
A1	160	j/a
A2	120	j/a
A3	40	j
A4	26	j
A5	(broken in shipment)	
A6	110	j/a
	Average = 91	

Set-B (T_{join}=1600°C, t_{join}=0.5 hr.)

Sample Number	MOR (MPa)	Path
B2	50	j
B4	253	j/a
B6	218	j/a
B8	124	j/a
B10	122	j/a
B12	124	j/a
	Average = 148	

Set-C (T_{join}=1600°C, t_{join}=0.5 hr., $T_{crystallize}$=1100°C, $t_{crystallize}$=1.0 hr.)

Sample Number	MOR (MPa)	Path
C2	193	j/a
C4	136	j/a
C6	97	j
C8	140	j/a
C10	159	j/a
C12	83	j/a
	Average = 135	

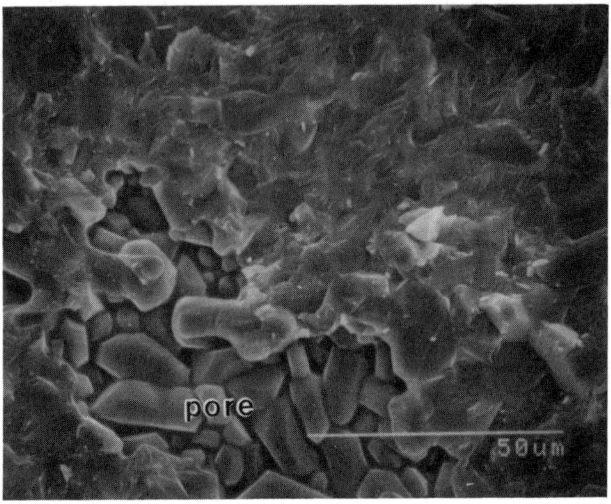

Figure 9. Secondary electron image of the fracture surface of a specimen that was machined from a sample formed using a 0.5 hr anneal at 1600°C and loaded to failure under four-point loading at room temperature.

Table III reports the results of mechanical testing of bars from Sets-B and -C performed at 500°C. The average strength displayed by the bars from Set-B was 144 MPa and 96 MPa for those from Set-C. These results also show wide scatter, standard deviations of 54 and 45 MPa.

Bulk samples of An-glass were analyzed in an effort to clarify the results obtained for the alumina joints. X-ray diffraction confirmed crystallization for both bulk An-glass samples and glazed alumina substrates given one hour anneals at 1100°C.

Small An-glass bars (1.5x2x20 mm) in both the vitreous and devitrified state were broken at room temperature in three-point bending. These samples were prepared by melting the glass in an alumina crucible ≈2.5 cm in diameter and ≈7.5 cm high. After melting the crucible was sacrificed to obtain the samples. The two glassy bars failed under stresses of 140 and 148 MPa, yielding a mean of 144 MPa, but the three crystalline bars broke under stresses of only 24.9, 23.2, and 46.0 MPa, yielding an average strength of 30.8 MPa.

An optical micrograph of the surface of a polished An-glass sample, ≈2 mm thick and ≈20 mm in diameter, which had been subjected to a crystallization anneal is presented in Figure 10. This surface shows that microcracking was prevalent in the sample. In addition, this annealed sample exhibited significant warpage presumably a result of stresses produced by crystallization.

Table III. Mechanical testing results using four-point bend method at 500°C on An-glass joints. The notation "j" indicates that the failure path was through the joint and "j/a" indicates that path at veered from the interlayer into the alumina.

Set-B (T_{join}=1600°C, t_{join}=0.5 hr.)

Sample Number	MOR (MPa)	Path
B1	71	j
B3	188	j/a
B5	199	j/a
B7	186	j/a
B9	121	j
B11	96	j/a
	Average = 144	

Set-C (T_{join}=1600°C, t_{join}=0.5 hr., $T_{crystallize}$=1100°C, $t_{crystallize}$=1.0 hr.)

Sample Number	MOR (MPa)	Path
C1	75	j/a
C3	124	j/a
C5	52	j
C7	114	j/a
C9	51	j
C11	162	j/a
	Average = 96	

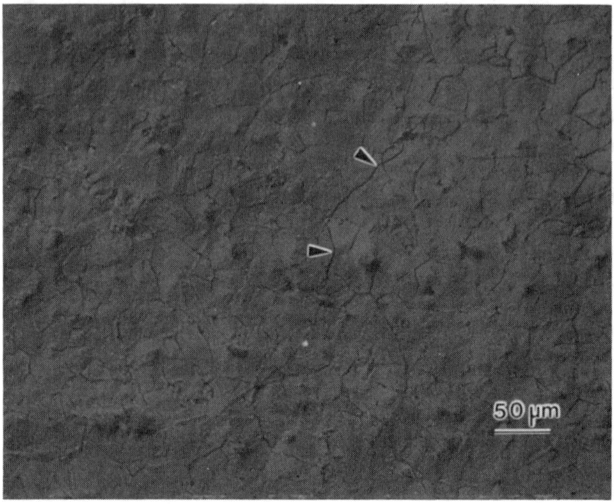

Figure 10. Optical micrograph (200 X, Nomarski imaging) of a bulk sample of An glass that had been polished using 1/4 µm diamond grit and then given a devitrification anneal for 1.0 hr at 1100°C. A high number density of crack developed during crystallization, the arrows point to one example.

DISCUSSION

An-glass was readily made by melting the appropriate batch and then allowing the melt to cool, i.e. the resultant glass was transparent, was free of unmelted batch, had a low density of bubbles and remained vitreous upon cooling. The latter quality allows the glass to be crystallized under controlled conditions at later point. Using a prereacted frit as the interlayer material avoids the need for homogenization during the joining step. An-glass was confirmed to crystallize after an anneal of 1 hr at 1100°C in both in its bulk form and in contact with an alumina substrate.

The microstructure of An-glass-bonded alumina joints displayed continuous, uncracked interfaces and low levels of porosity in the join region for samples held at 1600°C for 0.5 hr. These characteristics are essential for the formation of seals that are vacuum-tight and mechanically sound. Within regions where the interlayer was reasonably thick (on the order of 100 µm), acicular alumina precipitates were discovered. Penetration of the An-glass into the grain boundaries of the alumina was observed, which was consistent with prior work [20-22, 29]. Finally, cracking was observed in an interlayer that had been given a post-join anneal and in bulk specimens of interlayer material which had been devitrified.

The psuedobinary phase diagram [23] for anorthite and alumina suggests that alumina will be dissolved at the brazing temperature and reprecipitate upon cooling. This predicted precipitation was observed, but since a glass forms upon cooling of the joint, the microstructure necessarily represents a nonequilibrium distribution of phases. Therefore, somewhere between 4 and 10 percent alumina remained dissolved in the An-glass. From a calculation of the fraction occupied by the alumina crystals within the braze it is estimated that nearly 6 weight % alumina precipitated indicating that the final glass composition is very near that of the eutectic. Sample to sample variations in the EMPA data prevented it from being used to confirm this inference, however the whiskers were determined to be alumina.

A large fraction of these whiskers, 73%, were aligned at angles of 60° to 90° to the adherend faces. Such directional growth of crystals from the adherend surface has been viewed by others in studies on joining of alumina [24] and silicon nitride [30] and it has been inferred, though not proven, that this may result in an increased resistance to fracture.

Evidence for grain-boundary penetration by the interlayer glass was provided by measuring the interlayer thickness coupled with elemental analysis. The latter also indicate that the composition of the glass remains uniform as it moves away from the interface. Therefore anorthite should be produced throughout the microstructure during the crystallization anneal.

Penetration of the silicate braze into the grain boundaries of the alumina is significant for several reasons. Although the additional glass introduced into the alumina creates an alumina body with a lower overall alumina concentration than the original body, this concentration decrease should amount to no more than a few percent. Any decrease in the mechanical properties such as creep resistance of the adherend would be modest. If the glass is uniformly distributed throughout the alumina, the effects the glass could have on the mechanical properties are minimized. The penetration also results in a reduced interlayer thickness and in the formation of an interlocking glass network. Joint strength has been shown to increase with decreasing thickness by Zdaniewski and others [11]. Johnson and Rowcliffe [17] have further shown that there is an optimum. In the latter study the strength increased with decreasing initial thickness down to a starting thickness of approximately 30 µm. Below this thickness joint coverage sometimes became incomplete. In principle, brazing with An-glass allows the initial braze thickness to be relatively large so that joint coverage does not pose a problem and a thin final interlayer thickness may be obtained as a result of the glass migration. However, significant porosity was evident in metallographic specimens processed using long time anneals (2.0 hours at 1600°C). This suggests that some excess residual glass at the interface may be preferable for samples with a finite roughness.

Fractography indicates that coverage was incomplete despite the large initial thickness. Presumably this could be avoided by use of glass foils rather than a powdered glass suspension.

Upon crystallization of the An-glass interlayer substantial crack growth is visible within the braze. This cracking tends to be parallel to the adherend faces, consistent with the stress analysis performed by Kirchner and others [10].

This cracking could have been induced by stresses that developed during the volume change upon crystallization or may have arisen from the anisotropic thermal expansion of crystalline anorthite [25]. This anisotropy is believed to be large enough to be able to produce microcracking, based on similar occurrences in titania [26].

The latter explanation is consistent with the observation that the crystallized joints were always on average weaker, particularly at 500°C. Regardless of the explanation, the drastic decrease in the modulus of rupture observed when the bulk An-glass specimens were crystallized indicates that, in contrast to the solder glasses [4], devitrification should be avoided when this system is employed.

The room temperature strength of the uncrystallized and crystallized An-glass joints compare well with the strengths of alumina/alumina joints bonded with other interlayer compositions. Of the glass adhesives used by Zdaniewski and others to join alumina [11], the highest attainable average three-point bend strength was 90 MPa. Klomp and Botden [13, 14] reported three-point bend strengths of alumina with various compositions that they felt to be suitable for use at elevated temperatures. Compositions from the CaO-Al_2O_3-MgO-SiO_2 system [14] were used to form seals that had strengths ranging from 110 MPa to 220 MPa. A composition very similar to that used in this work, 42% CaO, 18% Al_2O_3, and 40% SiO_2 by weight, yielded a similar average bond strength of 150 MPa [14]. Klomp and Botden joined alumina using interlayers from the Al_2O_3-MnO-SiO_2 system [13] and produced a maximum average strength of 178 MPa. Brazing of the same type of alumina used in this current study with a filler metal having a composition of 45 Cu-26 Ag-29 Ti, in atomic percent, resulted in joints with an average four-point bend strength of 222 MPa at 25°C [8].

One study of interest was one in which silicon nitride was joined with a glass that was chosen to match the glassy grain-boundary phase and to crystallize during a post-join anneal [17]. This joining system is of particular interest because it displayed the same type of effects from a solution-precipitation reaction [17,30] that were observed when joining alumina with An-glass. Four-point bend strengths of these joints were as high as 460 MPa at room temperature [17]. It should be noted that these joints were produced using an intermediate glazing and grinding step which precluded porosity at the interface.

Although mechanical testing of joints targeted for high-temperature use needs to be performed at elevated temperature to assure their structural integrity, little high-temperature strength data is available for such joints. In this study, the vitreous joints retained at 500°C the strength that they exhibited at room temperature. These results are similar to those obtained for alumina brazed with Cu-Ag-Ti alloys. This system has been shown to possess a mean four-point bend strength in air of 215 MPa at 400°C and 213 MPa at 600°C, compared to a room temperature value of 222 MPa [8].

CONCLUSIONS

(1) An-glass was shown to meet all the original criteria for a silicate braze. It is a stable, high melting composition that wets and penetrates grain boundaries of alumina and can be crystallized through a post-formation anneal.

(2) An-glass proved to be easy to process: synthesis was straightforward; powder application was flexible and easy to optimize; and the fired joints had low levels of porosity.

(3) Vitreous joints were useful to at least 500°C and displayed superior mechanical properties at both temperatures. Average stresses of 148 MPa and 135 MPa were required to break specimens with uncrystallized and crystallized interlayers, respectively, at room temperature. Uncrystallized joints retained their room temperature strength at 500°C, while the average failure stress of devitrified joints dropped by 29 percent. Thus, in this system

crystallization degrades joint performance.

(4) Solution and precipitation reactions occur in the anorthite-alumina system during thermal cycling to 1600°C and lead to the penetration of the glass into the grain boundaries and the growth of alumina crystals into the interlayer of thick joints.

LIST OF REFERENCES

1. Ceramic Processing. 1968. Prepared by the Committee on Ceramic Processing, National Academy of Sciences, Washington, Publication 1576.

2. Rice, R. W. 1976. Joining of Ceramics. pp. 69-111 in Advances in Joining Technology. J. J. Burke, A. E. Gorum, and A. Tarpinian, eds. Boston: Brook Hill.

3. Cawley, J. D. 1989. Joining of Ceramic Matrix Composites. Am. Ceram. Soc. Bull. 68 (9) : 1619-1623.

4. Francel, J. 1986. Sealing Glasses," pp. 79-86 in Advances in Ceramics, Vol. 18. D. C. Boyd and J. F. MacDowell, eds. Columbus, OH: American Ceramic Society.

5. Hauth, W. E. 1982. Joining of Technical Ceramics. CFI / Ber. DKG **59** (1) : 12-16.

6. Mizuhara, H., and Huebel, E. 1986. Joining Ceramic to Metal with Ductile Active Filler Metal. Welding Journal **65** (10) : 43-51.

7. Gitzen, W. H. 1970. Seals, Metallizing, Welding, pp. 189-192 in Alumina as a Ceramic Material. Columbus OH : American Ceramic Society.

8. Moorhead, A. J. 1987. Direct Brazing of Alumina Ceramics. Adv. Ceram. Mats., **2** (2) : 159-166.

9. Borom, M. P. 1978. The Mechanical and Chemical Aspects of Glass Sealing. The Glass Industry **59** (3) : 12-16.

10. Kirchner, H. P., Conway, J. C., Jr., and Segall, A. E. 1987. Effect of Joint Thickness and Residual Stresses on the Properties of Ceramic Adhesive Joints: I, Finite Element Analysis of Stresses in Joints. J. Am. Ceram. Soc., **70** (2) : 104-109.

11. Zdaniewski, W. A., Conway, J. C., Jr., and Kirchner, H. P. 1987. Effect of Joint Thickness and Residual Stresses on the Properties of Ceramic Adhesive Joints: II, Experimental Results. J. Am. Ceram. Soc., **70** (2) : 110-118.

12. Hauth, W. E., III 1979. Ceramic-to-Ceramic Sealing of Large Shapes. Am. Ceram. Soc. Bull. **58** (6) : 584-586.

13. Klomp, J. T., and Botden, Th. P. J. 1970. Sealing Pure Alumina Ceramics to Metals. Am. Ceram. Soc. Bull. **49** (2) : 204-211.

14. Klomp, J. T., and Botden, Th. P. J. 1965. A New Sealing Method for High-purity Alumina. pp. 129-139 in "The Use of Ceramics in Valves," Proc. Symp. Brit. Cer. Res. Assoc., Spec. Publ. No. 48.

15. Zdaniewski, W. A., Shah, P. M., and Kirchner, H. P. 1987. Crystallization Toughening of Ceramic Adhesives for Joining Alumina. Adv. Ceram. Mats. **2** (3A) : 204-208.

16. Clark, T. J., and Reed, J. S. 1986. A Novel Technique for Producing a Glass-Ceramic Bond. Am. Ceram. Soc. Bull. **65** (11) 1506-1512.

17. Johnson, S. M., and Rowcliffe, D. J. 1985. Mechanical Properties of Joined Silicon Nitride. J. Am. Ceram. Soc. **68** (9) : 468-472.

18. Cawley, J. D., and Knapp, J. T. 1990. Silicate Brazing of Alumina Ceramics. Report B9001, Edison Welding Institute, Columbus, OH.

19. Dogan, C. 1989. Ph. D. dissertation, Case Western Reserve University, Cleveland, OH.

20. Huppman, W. J., Pejovnik, S., and Han, S. M. 1978. Rearrangement During Liquid Phase Sintering of Ceramics. pp. 233-242 in Materials Science Research, Vol. 11. Palmour, H., III, Davis, R. F., and Hare, T. M. New York: Plenum Press.

21. Flaitz, P. L., and Pask, J. A. 1987. Penetration of Polycrystalline Alumina by Glass at High Temperatures. J. Am. Ceram. Soc. **70** (7) 449-455.

22. Shaw, T. M., Cook, R. F., and Duncombe, P. R. 1990. The Kinetics of Grain Boundary Wetting in High Purity Alumina. Technical presentation, 92nd annual meeting of American Ceramic Society, Dallas, TX.

23. Phase Diagrams for Ceramists. Columbus, OH: American Ceramic Society.

24. Jonker, G. H., Klomp, J. T.,and Botden, Th. P. J. 1964. High-temperature Seals on Pure Dense Alumina. pp. 295-302 in Science of Ceramics, Vol. 2. G. H. Stewart, ed. New York: Academic Press.

25. Kozu, S., and Ueda, J. 1933. Thermal Expansion of Plagioclase. Proc. Imp. Acad. (Tokyo) **9**, 262.

26. Kingery, W. D., Bowen, H. K., and Uhlmann, D. R. 1976. Introduction to Ceramics, Second Edition. New York : Wiley-Interscience.

27. Taylor, J. R., and Bull, A. C. 1986. Ceramics Glaze Technology. New York: Pergamon Press.

28. Towers, T., and Chipman, J. 1957. Diffusion of Calcium and Silicon in a Lime-Alumina-Silica Slag. Trans. AIME 209 (June) : 769-773.

29. Shaw, T. M. August 3, 1990. IBM Thomas J. Watson Research Center. Yorktown Heights NY.

30. Mecartney, M. L., Sinclair, R., and Loehman, R. E. 1985. Silicon Nitride Joining. J. Am. Ceram. Soc. 68 (9) : 472-478.

Electronic Applications

MEASUREMENTS OF PROPERTIES OF MATERIALS IN MULTILAYER CERAMIC CAPACITORS AND MICROELECTRONIC PACKAGING

Joseph A. Carpenter, Jr.
Ceramics Division
National Institute of Standards and Technology
Gaithersburg, MD 20899

Abstract

There are needs for the abilities to measure the electrical, mechanical, thermal and chemical state properties of the materials at and near interfaces in multilayer ceramic capacitors (MLCCs) and in microelectronic packaging. Knowledge of the absolute values of such properties is important for their design and applications. In the case of MLCCs, the needs are primarily associated with the continuing problem of in-service mechanical failures, while in the case of microelectronic packaging, they are associated with the trend toward denser features. Techniques for such measurements are not widely used, particularly for the micrometer-scale and most especially those suited for use on-line in an industrial setting. What techniques are available are covered and possibilities for future developments are discussed.

Metal - Ceramic Joining
Edited by P. Kumar and V. A. Greenhut
The Minerals, Metals & Materials Society, 1991

Introduction

Multilayer ceramic capacitors (MLCCs) and packaging, ubiquitous in modern microelectronics based on microchips, are components consisting of thin layers of conductors embedded in thicker layers of dielectrics. They are required to perform reliably for a reasonable cost. The objectives of high reliability and low cost are largely achieved today through computerized design using bulk material properties and extensive product testing. As illustrated in Figure 1, the features of such components are being forced to downsize into micrometer sizes as the features on the chips decrease into sub-micrometer sizes in order to achieve faster speeds and higher levels of integration. At such fine scale the assumption of bulk property values becomes suspect, especially in areas near the interfaces, both internal and external.

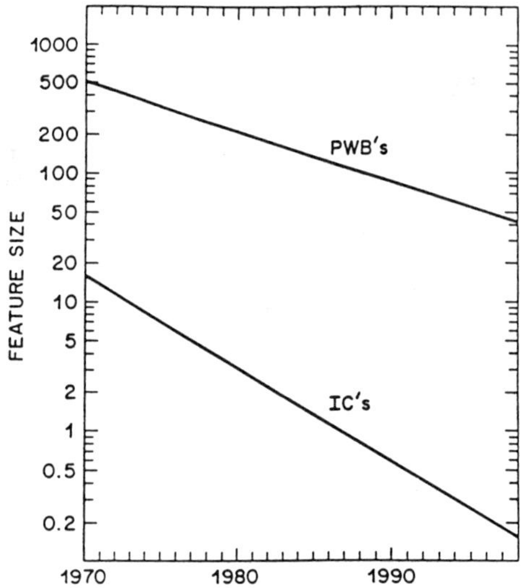

Figure 1. The sizes of the features in multilayer ceramic capacitors (MLCCs) and packaging for microelectronic components are being forced into the micrometer scale range by the downsizing of the features on microchips into the sub-micrometer range in order to increase speed and level of integration. In this figure, typical sizes, in micrometers, of the features in printed wiring boards (PWBs) are compared with those on the integrated circuits (ICs) as functions of years. (Source: Ref. 1, page 17)

This paper addresses the results of recent interactions by NIST with industrial, academic, and government personnel who have identified material property measurements of most interest for MLCCs and packaging for microelectronics and suggests ways of making such measurements. Knowledge of the absolute values of such properties is important for the design and applications of the MLCCs and packaging. Desires have been expressed for development of capabilities for measuring electrical, thermal, mechanical and chemical state properties of the materials as actually found in MLCCs and microelectronics packaging, not just as measured on bulk specimens. Today, these four properties of interest are most often assumed by comparing images and elemental analyses of microstructures of MLCCs and packaging with images and elemental analyses of microstructures of specially prepared bulk specimens whose properties have been measured by macroscale methods. Such capabilities would find use in research and development, product design, manufacturing process control and quality checks, in-service diagnostics, and failure analyses.

Because of the very broad nature of the subject matter and limitations in space, only brief mention of the various approaches for making such measurements is made and no discussions of expected accuracy and precision are possible. Current references are given from which other more detailed references can be found. In general, techniques for measuring such properties at the fine scale of the features in the MLCCs and packaging have either not been developed or applied to determine their limits of accuracy and precision or applicability to research and development, design, manufacturing, in-service diagnostics, or failure analyses. Techniques for microstructural imaging and element identification and imaging at the micrometer-size scale of the features in such components are readily available (e.g., see Refs. 2 and 3) and thus will not be addressed.

Multilayer Ceramic Capacitors

Longitudinal cross sections and typical defects in MLCCs are shown in Figures 2 and 3. These devices and their manufacture have been described elsewhere (4,5). The MLCC consists of thin films of metal electrodes, typically a few micrometers thick and nearly as long and as wide as the capacitor (typically a few millimeters), sandwiched between layers of ceramic dielectric, typically tens of micrometers thick. The electrode metals are usually Ag and its alloys with Pd, Pt, and Au, mixed with glass and other proprietary additives to bond the metals to the ceramic. The ceramic dielectrics are various oxides with barium titanate ($BaTiO_3$) being the most prevalent. The ceramic powders are also mixed with glass and other additives to allow firing temperatures that the metals can withstand. The ends are usually terminated with a layer of Ag or Ag-Pd paint, which is plated with a layer of nickel, which is covered with solder; the silver paint serves to connect the

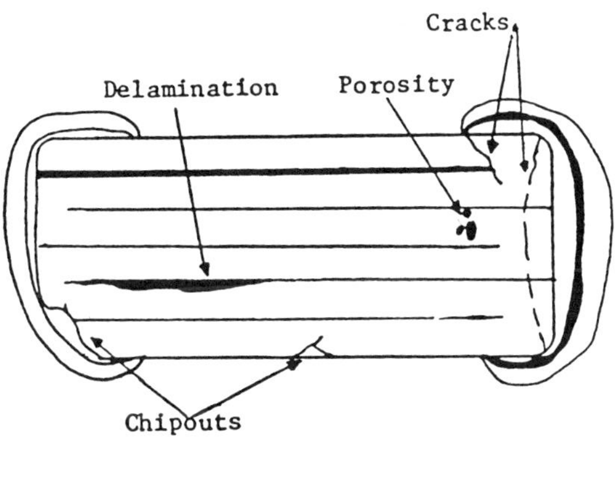

Figure 2. Cross section of a multilayer ceramic capacitor showing typical manufacturing defects. (Source: Ref. 6)

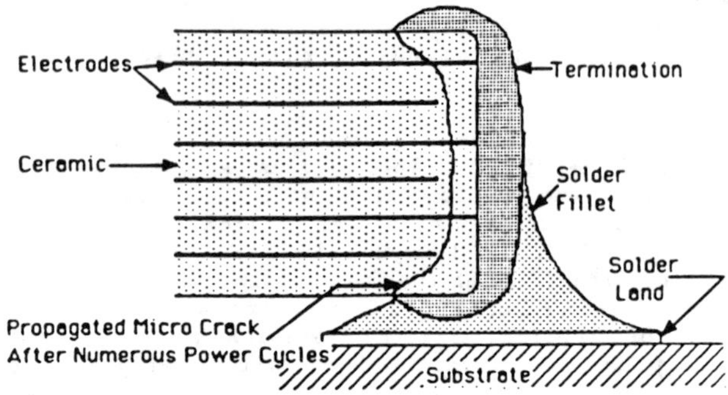

Figure 3. Cross section of an MLCC soldered to a substrate via surface mount technology (SMT) illustrating a typical cracking pattern due to power cycling. (Source: Ref. 7)

electrodes and provide an electrical contact for electroplating the nickel. The nickel serves as a barrier to diffusion of the Sn from the solder into the body of the MLCC. The final solder covering ensures wetting of additional solder used to connect the MLCC to external leads.

Currently, the main issues with MLCCs are reliability and cost. While only a few years ago in-service failure rates of 2,500 per million were acceptable, today they are on the order of a few per million (8) and demand for a few per billion is expected. As most of these components sell for only a few cents, the challenge is to achieve such low rates of failures at reasonable costs.

Most in-service failures are associated with defects such as porosity, delaminations and cracks introduced during manufacturing, component assembly, or in-service. Today's low in-service failure rates have been achieved through various means, notably improvements in manufacture to reduce the porosity and delaminations, extensive screening by the manufacturer prior to shipment (which imposes additional costs on the manufacturer), and greater care on the part of component assemblers; however, the cracking problem remains.

Several causes of cracking have been identified (7). Among the most common is mechanical stress thermally induced by wave soldering (5,9) or power cycling of MLCCs soldered directly onto the boards (6) using surface mount technology (SMT). Typical cracking patterns resulting from mounting and subsequent power cycling in a surface mounted MLCC are depicted in Figure 3. This results from the differential thermal expansions of the substrate, the solder and the MLCC. Cracks typically originate at the corners of the shoulders of the end terminations and propagate inward. Modeling and experimental studies (6, 9-12) have identified thermal diffusivity, thermal coefficient of expansion (TCE), stress-strain responses (Young's modulus, shear modulus, Poisson's ratio) fracture strength and fracture toughness as the most important material properties affecting cracking and, therefore, should be measured. In addition, residual stresses introduced during manufacture, assembly, or in service are also suspected to play a role in the cracking and thus techniques to detect their presence are also desired.

Packaging

The primary functions of packaging are to transmit power to the chip, transmit signals to and from the chip, dissipate the heat generated by the chip, and provide it a survivable environment. The trend toward ever increasing speeds demands packaging dielectrics with dielectric constants and loss factors that are as low as possible. The trend toward increased levels of integration of functions on the chips implies larger chips with finer features and higher heat outputs. The higher level of heat output requires packaging with reliable thermal properties and the ability to withstand thermally induced mechanical stresses. The packaging materials themselves must be chemically stable in order to achieve reliable properties over time. This technology has been thoroughly described in recent works (13-15).

Though there are a myriad of packages available today, the important current issues can be identified by reference to Figure 4, in which a thin film (16) polymer-on-cofired-ceramic multichip package is depicted in cross section. Metallization on the chip is electrically connected through solder bumps to metallization on the upper surface of the package. The upper part of the package consists of conductor stripes, sometimes made of aluminum but more usually of chromium- or nickel-coated copper, which are typically a few micrometers thick and a few tens of micrometers wide, sandwiched between dielectric layers of polymer, mostly polyimide but sometimes benzocyclobutene, which are typically a few tens of micrometers thick. The lower structure consists of conductors, typically made of tungsten, molybdenum, copper or gold, which are a few tens of micrometers thick and about one or two hundred micrometers wide sandwiched between dielectric layers, mostly alumina containing some glass, about a hundred micrometers or more thick. Conductors penetrating though the upper and lower parts connect to pins which connect to the next level of packaging, typically a printed wiring board (PWB).

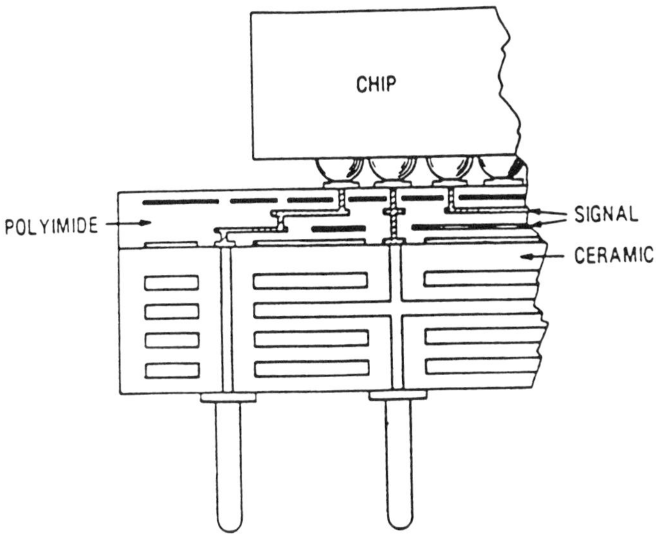

Figure 4. Cross section of thin film polymer-on-cofired-ceramic multichip package. (Source: Ref. 1, page 130)

The needs perceived for measuring the properties of materials in packaging were identified in a workshop (17) held at NIST in May 1990. Of most interest are techniques for measuring the dielectric properties, the thermal diffusivity, thermal coefficient of expansion, stress-strain responses, adhesion, and fracture toughness and for detecting the presence of residual stress and the degree of cure of the polymers.

Measurement Techniques

Electrical Properties

Measurements of electrical properties are of current interest in packages. The first function of the package is to supply power and ground and transmit signals to and from the chip. Most of the concern today is with the latter. In modern high performance digital computers, for example, the signal is a voltage pulse that must propagate from a function on one chip, called a driver, to a function on another chip, called a receiver, in times on the order of nanoseconds. Furthermore, the signal must arrive at the receiver sufficiently undistorted. The controlling properties are the resistance of the conductor, magnetic permeability of the conductor and dielectric, and dielectric constant and loss factor of the dielectric; of these, the dielectric constant and loss factor are of most concern because they can be the most variable. Increasing dielectric constant causes increasing signal propagation delay while the loss factor degrades the shape of the pulse. As pulse rise times lower than a hundred picoseconds are becoming commonplace, measurements of the dielectric constant and loss factor at frequencies up to about 100 gigahertz are needed.

There are three general methods for determining the dielectric constant and loss factor of bulk specimens: lumped element, resonant cavity, and transmission line (18). In the typical lumped element technique the sample is inserted between and in contact with two flat plate or coaxial electrodes of known dimensions and the capacitance and conductance measured to yield the dielectric constant and loss factor. Such methods can be used for measurements over a range of frequencies up to about 1 GHz. Resonant cavity and transmission line methods are required at frequencies above 1 GHz and are considerably more difficult than the lumped element techniques. In a resonant cavity method, the specimen is loaded into a resonant cavity and the changes in resonant frequency and quality factor with respect to the unloaded condition are measured to yield the dielectric constant and loss factor. In a transmission line method, changes of the amplitude, phase and/or shapes of waves propagated through or reflected by a sample are measured. Of these three approaches, only the lumped element and transmission lines would appear to have potential for measurement of the features in packages as it is difficult to see how a specimen could be obtained for loading into a resonant cavity.

Another possibility is the use of pulse propagation delay and shape analyses. In these approaches, the time delay and degradation of a pulse along a conductor surrounded by an insulator might be used to measure the dielectric constant and loss factor of the insulator. This approach is similar to time domain reflectometry commonly used in industry for assessing the electrical quality of packaging. These approaches also offer the possibility of using the conductors in the packaging for *in situ* measurements.

Techniques based on electron or laser beams are possibilities that may be explored. The electron beam techniques would utilize the decay of charge built up by exposure to the electron beam (similar to voltage contrast electron beam methods) whereas the laser techniques would measure the refractive index and deduce the dielectric constant through knowledge of its relationship to the refractive index. Both approaches would offer high spatial resolutions in directions perpendicular to the beams.

Thermal Properties

For both the MLCCs and packaging, the thermal diffusivity and the thermal coefficient of expansion are of interest as they determine the extent of the thermally induced stresses. The thermal diffusivity is the major material property determining the ability of the packaging to dissipate heat from the chip. Currently, the overall thermal resistances of packages are experimentally assessed by means of thermal test chips or other devices of known heat output mounted in the package and measurements of the overall temperature drop between the test chip or heating device and the exterior of the package. The temperature profiles in the package are then calculated via models using assumed bulk thermal properties (19-21).

Thermal Diffusivity. For bulk specimens, thermal diffusivity is usually calculated from the steady state thermal conductivity measured by the guarded hot plate or longitudinal bar methods (22) and values of the density and specific heat; neither approach appears to be directly applicable to the micrometer scale of the features in current and future MLCCs or packages.

Three photothermal methods, namely, thermal flash (or pulse), forced Rayleigh scattering, and thermal wave, would appear to have some potential. If the heating is supplied by a laser or electron beam with a micrometer-sized beam diameter, micrometer-scale spatial resolution in directions perpendicular to the beam is possible. Thermal flash or pulse methods (23) introduce a thermal pulse into one face of a flat specimen and the resultant thermal wave on the other side is detected by some means, usually a transducer. In forced Rayleigh scattering (24), two laser beams are crossed at the surface to create fringes which produce a standing heat wave pattern which serves as a temporary diffraction grating as result of the change of refractive index with temperature. A third laser beam is directed through this grating and the change of the diffraction angle with time after the two crossed beams are turned off yields the thermal diffusivity. Both methods would appear to be limited to free standing specimens specially prepared (and, therefore, not truly representative of actual MLCCs or packaging) or somehow sectioned from MLCCs or packaging without substantial alteration of the properties of interest. In thermal wave techniques (23,25) a thermal pulse or wave is introduced into a point on the surface of a specimen and the result at a point on the same surface is measured either by its effect on the temperature of the surface or the atmosphere just above it. It has the advantage that access to only one free surface is required.

Thin film thermocouples (26) appear to offer promising possibilities for determining thermal diffusivity and possibly thermal conductivity at the micrometer scale. Such thermocouples sputter deposited onto a substrate surface may be pulsed and the drop-off in temperature with time used to determine the diffusivity of the substrate. Since metals such as copper and platinum, used as conductors in MLCCs and packaging, are also used on thermocouples, the obvious possibility is that thin film thermocouples might be formed within the bodies of the MLCC and packaging multilayer structures thus providing actual measurements to verify the thermal models used today or monitor process and service conditions.

Thermal Coefficient of Expansion. Thermal coefficient of expansion (TCE) of bulk specimens is measured in various ways including dilatometry, telemicroscopy, interferometry, capacitance, and, for crystalline materials, x-ray diffraction (27). Dilatometry has been used for measuring the aggregate TCEs of MLCCs (e.g., 6). A capacitance technique has been recently described for measuring the TCE of thin films in the direction of film thickness (28). The use of laser-based interferometric techniques for measuring the TCE of supported films in the direction of the film thickness has also been discussed (29). Since MLCCs and multilayered packaging are laminated structures, means for also measuring TCEs in directions parallel to the planes of the layers are important; Moire interferometry might be a possibility. Resistive thin films used as strain gages are another possibility provided the effects of temperature on the resistance of the films can be removed; films with low thermal coefficients of resistance (TCRs) appear most promising.

Mechanical Properties

The mechanical properties are important as they determine the ability of the MLCC or packaging to resist stresses imposed thermally or mechanically during processing or in service. The main needs identified are the capabilities for measuring the stress-strain properties, strengths of adhesion, resistance to fracture, and residual stress.

Stress-Strain Responses. The main stress-strain response properties of interest are the Young's modulus in tension, the shear modulus and Poisson's ratio. These properties are measured in bulk specimens in a variety of ways including tensile testing, shear testing, flexure testing, ultrasonic testing, microindentation, and strain gaging. A technique for tensile testing of thin films using micromechanical gripping and strain gaging is under development at NIST (30). Flexure and ultrasonic testing have been used to measure the gross stress-strain properties of whole MLCCs (6) and supported thin films (31), but it is difficult to see how these approaches could be applied to the size scale of the features in the MLCCs and packaging. Microindentation has been used (32) to determine Young's modulus at high spatial resolution in MLCCs and would appear to have great appeal for future use. Miniature strain gages have been used to measure strains on the exteriors of MLCCs and packages (e.g., 10, 33). Thin film strain gages would seem to offer possibilities for direct measurements of the interior strains.

Adhesion. Of the myriad of adhesion tests available, the ones mostly commonly used in microelectronics include the direct pull-off, direct shear, peel test, scratch test, density-of-cracks, twist-off, lap shear, double cantilever, and microindentation (e.g., 34 and references therein). They all involve complicated stress states and yield measurements dependent in complicated ways on the material properties of the two adherends as well as the interface. They are best categorized as semi-quantitative screening tests rather than measurements of basic material properties. Use of such approaches at the scale of the features in MLCCs and packaging seems problematical and will require additional thought and insight. A major problem is quantitatively separating the strengths of the adherends from the strength of the interface. A possible method for measuring the interfacial strength is the recently developed surface forces apparatus (35) in which atomic forces between interfaces can be measured.

Fracture Resistance. Resistance to fracture is primarily a concern to the MLCCs as discussed above; it is often stated that ceramic packages simply do not fail (e.g., 13, p. 4). The material property of interest is fracture toughness. Fracture toughness of bulk ceramic specimens is measured by means of three- or four-point bending and by microindentation. Three- or four-point bending would not appear to be applicable to MLCCs without some means for extracting without damage thin films from the bodies as mentioned above in the discussion on thermal diffusivity. Microindentation has been used for studies of MLCCs (32,36).

Residual Stress. Residual stress is usually measured in bulk polycrystalline specimens by x-ray diffraction. Because of lack of intensity obtainable from normal x-ray sources, the area of incidence of the x-ray beam must be larger than would desirable for the fine scale of the features in MLCCs and packaging; a possibility for research and development might be synchrotron-based techniques employing two or more collimating crystals. Techniques based on shifts of spectral lines with stress may be applicable to MLCCs and packaging. Laser-based methods using infrared or Raman lines have been used to measure stress in surfaces of ceramics and polymers (37). Laser-based methods involving the shifts of fluorescence lines from materials deliberately salted into specimens has been in long use for measuring stress in materials under very high pressures. Electron beam-induced cathodoluminescence lines have also been shown to exhibit shifts with stress.

Chemical Properties

As stated above, there are a variety of techniques currently available for identifying and mapping the distributions of elements at the scale of the features in MLCCs and packages. While identification of the elements present is certainly important, knowledge of their chemical state is equally if not more important. The degree of cure of the polyimide and benzocyclobutene polymers in the packages has been identified as of most interest at the current time. The degree of cure is important as it determines the desired achievement of many of the other properties discussed here and the degree of stability of the polymer itself. For example, incomplete cure can leave unreacted chemicals which in the presence of moisture can corrode the copper conductors.

Techniques for possible use in polymeric matrix composites intended for structural applications have been identified (38) but to date none has been used in packaging. Dielectric properties are sensitive to molecular structure and mobility and in fact miniature dielectric sensors have been developed and used to monitor cure. Other possible approaches include fluorescence, ultraviolet absorption, infrared and Raman spectroscopy, and refractive index changes. The exciting radiation could be supplied to a free surface by means of a laser or to the interior by means of embedded optical fibers. Optical fibers used to date have been about ten micrometers in diameter, so questions remain as to their applicability in layers that are only tens of micrometers thick.

Summary

Capabilities for measuring the following are currently perceived as most desirable for the materials in multilayer ceramic capacitors and packaging for microelectronics:

(1) dielectric constant and loss factor from 0 to 100 GHz;
(2) thermal diffusivity;
(3) thermal coefficient of expansion;
(4) stress-strain response;
(5) adhesion;
(6) fracture toughness;
(7) residual stress;
(8) the degree of cure of polymers.

Most techniques for measuring such properties at the fine scale of the features in the MLCCs and packages are yet to be developed and assessed to determine limits of accuracy and precision and their applicability to research and development, process control and quality assessment in manufacturing, in-service diagnostics, and failure analyses. The development should be connected with modeling to determine what parameters are most important to measure and with what accuracy and precision. These techniques may allow studies to determine when the assumption of bulk properties in design is no longer valid.

Acknowledgements

Credit for identification of needs for measurements in packaging go to the participants, from inside and outside of NIST, in the workshop (17) held at NIST in May 1990. Thanks go to the various staff members of NIST who proposed ideas as to how the recommendations of the workshop might be addressed. Special thanks go to S. J. Dapkunas, S. W. Freiman, K. G. Kreider, and L. H. Schwartz of NIST who reviewed and critiqued this manuscript.

References

1. D. W. McCall, chairman, Materials for High-Density Electronic Packaging and Interconnection, Report of the Committee on Materials for High-Density Electronic Packaging, National Materials Advisory Board, Commission on Engineering and Technical Systems, the National Research Council (Washington, DC: NMAB-449, National Academy Press, 1990).

2. Eugene S. Meieran, Paul A. Flinn, and John R. Carruthers, "Analysis Technology for VLSI Fabrication," Proceedings of the IEEE, 75 (7) (1987), 908-955.

3. Lawrence Kashar, "Microanalytical Techniques: Their Use in Solving Problems in the Manufacturing and Reliability of Electronic Packages, Electronics Packaging Forum, Volume One, ed. J. E. Morris (New York, NY: Van Nostrand Reinhold, 1990), 219-281.

4. Gilbert Goodman, "Ceramic Capacitor Materials," Ceramic Materials for Electronics, ed. R. C. Buchanan (New York, NY: Marcel Dekker, Inc., 1986), 98-103.

5. Bharat S. Rawal, Richard Ladew, and Ricardo Garcia, "Factors Responsible for Thermal Shock Behavior of Chip Capacitors," Proceedings of 37th Electronic Components Conference (New York, NY: Institute of Electrical and Electronic Engineers, 1987), 145-156.

6. Michael J. Cozzolino, "Stress Determination for Surface Mounted Ceramic Capacitors," Proceedings of 6th Capacitor and Resistor Technology Symposium (CARTS) (Huntsville, AL: The Components Technology Institute, Inc., 1986), 42-52.

7. John Maxwell, "Cracks: The Hidden Defect," Proceedings of 9th Capacitor and Resistor Technology Symposium (CARTS) (Huntsville, AL: The Components Technology Institute, Inc., 1989), 47-54.

8. T. J. Dwyer and R. B. McPhillips, "Growth and Competition in the World Electronic Ceramics Market," USACA Report (newsletter of the United States Advanced Ceramics Association, Suite 300, 1440 New York Avenue, NW, Washington, D.C. 20005, 202-638-1200, fourth quarter 1988), 3.

9. Bharat S. Rawal et al., "Reliability of Multilayer Ceramic Capacitors After Thermal Shock," Proceedings of 38th Electronic Components Conference (New York, NY: Institute of Electrical and Electronic Engineers, 1988), 371-375.

10. Michael J. Cozzolino and Gary J. Ewell, "A Fracture Mechanics Approach to Structural Reliability of Ceramic Capacitors," IEEE Transactions on Components, Hybrids, and Manufacturing Technology, CHMT-3 (2) (1980), 250-257.

11. Gary J. Ewell, "Critical Mechanical Properties of Ceramic Chip Capacitors," Proceedings of 2nd European Capacitor and Resistor Technology Symposium and Seminar (Huntsville, AL: The Components Technology Institute, Inc., 1986), 60-70.

12. Grant Marten and Anita Danford, "Thermal Cracking of Ceramic Capacitors," Proceedings of 9th Capacitor and Resistor Technology Symposium (CARTS) (Huntsville, AL: The Components Technology Institute, Inc., 1989), 31-36.

13. Rao R. Tummala and Eugene J. Rymaszewski, eds., Microelectronics Packaging Handbook (New York, NY: Van Nostrand Reinhold, 1989).

14. Merrill L. Minges, technical chairman, Electronic Materials Handbook, Vol. 1, Packaging (Metals Park, OH: ASM International, 1989).

15. David P. Seraphim, Ronald C. Lasky, and Che-Yu Li, eds., Principles of Electronic Packaging, Design and Materials Science (New York, NY: McGraw-Hill Book Company, 1989).

16. There is currently some ambiguity in the electronics literature to the term "thin film." While chip technologists usually mean layers with thicknesses from a few tenths of nanometers up to a few micrometers, packaging technologists usually mean layers with thicknesses on the order of a few to a few tens of micrometers. It is the latter size that is meant by references to thin films in this paper. The common factor in the use of the term "thin film" by both the chip and packaging technologist seems to be that similar techniques such as chemical vapor deposition or sputtering are used to fabricate the conducting and dielectric layers rather than "thick film" fabrication techniques typically involving slurries and powders that produce layers with thicknesses from a few ten to a few hundreds of micrometers. For a good discussion see pages 673-676 of reference 13.

17. Joseph A. Carpenter, Jr., Workshop Coordinator, "Measurements of the Properties of Materials in Microelectronics Packaging," (Report to be published, National Institute of Standards and Technology, 1991).

18. Joseph A. Carpenter, Jr., "Dielectric Properties Measurements and Data," Microwave Processing of Materials, ed. W. B. Snyder, W. H. Sutton, D. L. Johnson, and M. F. Iskander (Pittsburgh, PA: Materials Research Society, 1991), to be published.

19. Frank F. Oettinger and Davis L. Blackburn, "Thermal Resistance Measurements," (NIST Special Publication 400-86, National Institute of Standards and Technology, July 1990).

20. R. H. Jensen et al. "Comparative Thermal Performance of Various Substrate Materials in a Simple Packaging Application: Actual vs. Predicted," Proceedings of 39th Electronic Components Conference (New York, NY: Institute of Electrical and Electronic Engineers, 1989), 572-576.

21. William E. Pence and J. Peter Krusius, "Package Thermal Resistance: Geometrical Effects in Conventional and Hybrid Packages," IEEE Transactions on Components, Hybrids, and Manufacturing Technology, CHMT-13 (2) (1990), 225-251.

22. E. S. Dettmer et al., "Steady State Thermal Conductivity Measurements of AlN and SiC Substrate Materials," Proceedings of 39th Electronic Components Conference (New York, NY: Institute of Electrical and Electronic Engineers, 1989), 551-556.

23. H. Coufal, "Photothermal Analysis of Thin Polymer Films," New Characterization Techniques for Thin Polymer Films, ed. H.-M. Tong and L. T. Nguyen (New York, NY: John Wiley and Sons, Inc. 1990), 231-264.

24. Qui Tran-Cong, Taihyun Chang, and Charles C. Han, "Polymers Being Intramolecular Photodimerizable Probes for Mass Diffusion Measurements by the Forced Rayleigh Scattering Technique," Polymer 29 (1988), 2261-2270.

25. George Birnbaum and Grady S. White, "Laser Techniques in NDE," Nondestructive Testing, Vol. 7 (London: Academic Press, 1984), 283-291 and 334-353.

26. D. Burgess, M. Yust and K. G. Kreider, "Thin Film Thermocouples for Measurement of Wall Temperatures in Internal Combustion Engines," Heat Transfer in Combustion Systems - 1990, ed. B. Farouk et al. (New York, NY: The American Society of Mechanical Engineers, 1990), 43-48.

27. Y. S. Touloukian et al., Thermophysical Properties of Matter, Vol. 13: Thermal Expansion, Nonmetallic Solids (New York, NY:Plenum Publishing Corporation, 1977), 17a-26a.

28. Ho-Ming Tong and Ravi F. Saraf, "Anisotropy in Coefficient of Thermal Expansion of Planar Oriented Polyimide Films," (Paper presented at the 1990 Fall Meeting of the Materials Research Society, Boston, MA, 28 November 1990).

29. K. L. Saenger and Ho-Ming Tong, "Laser Interferometry: A Measurement Tool for Thin Film Polymer Properties and Processing Characteristics," New Characterization Techniques for Thin Polymer Films, ed. H.-M. Tong and L. T. Nguyen (New York, NY: John Wiley and Sons, Inc., 1990), 117-119.

30. David T. Read, "A Microfabrication Approach to Mechanical Testing of Thin Films," (Paper presented at the 1990 Fall Meeting of the Materials Research Society, Boston, MA, 27 November 1990).

31. H.-M. Tong and K. L. Saenger, "Bending-Beam Characterization of Thin Polymer Films," New Characterization Techniques for Thin Polymer Films, ed. H.-M. Tong and L. T. Nguyen (New York, NY: John Wiley and Sons, Inc., 1990), 29-56.

32. B. Wong et al., "Microindentation for Mechanical Assessment of Microelectronic Components," Proceedings of the 1986 International Symposium on Microelectronics (Reston, VA: The International Society for Hybrid Microelectronics, 1986), 399-410.

33. Peter M. Hall, "Strain Measurements During Thermal Chamber Cycling of Leadless Ceramic Chip Carriers Soldered to Printed Boards" Proceedings of 34th Electronic Components Conference (New York, NY: Institute of Electrical and Electronic Engineers, 1984), 107-116.

34. Wolfgang Sachse, Kwang Yul Kim, and H. D. Conway, "Determination of the Bond Strength of a Metal Coating on a Substrate," Electronic Packaging Materials Science IV, eds. Ralph Jacodine, Kenneth A. Jackson, Edwin D. Lillie and Robert C. Sundahl (Pittsburgh, PA: Materials Research Society, 1989), 293-303.

35. Jacob N. Israelachvili and Patricia M. McGuiggan, "Adhesion and Short-Range Forces Between Surfaces. Part I: New Apparatus for Surface Force Measurements," Journal of Materials Research, 5 (10) (1990), 2223-2243.

36. Theresa L. Baker and Stephen W. Freiman, "Fracture Behavior of Ceramics Used in Multilayer Capacitors" Electronic Packaging Materials Science II ed. K. A. Jackson, R. C. Pohanka, D. R. Uhlmann, and D. R. Ulrich (Pittsburgh, PA: Materials Research Society, 1986), 81-90.

37. James W. Rydzak and Roger W. Cannon, "Fourier Transform Infrared Microspectroscopy Method to Determine Stress in Sapphire," Journal of the American Ceramic Society, 72 (8) (1989), 1559-1561.

38. Donald Hunston et al., "Assessment of the State-of-the-Art for Cure Sensors for Composite Materials," (Report by the Polymers Division of the National Institute of Standards and Technology, Gaithersburg, Maryland, to the National Center for Manufacturing Science, to be published as a NIST Report in 1991).

GRADED METAL-CERAMIC MICROJOINTS IN PARALLEL

Chou H. Li

Lintel Technology, Inc.
379 Elm Drive
Roslyn, NY 11576

Abstract

A basic worldwide problem in metal-ceramic joining exists with the mismatch in the thermal expansion coefficient between metal and ceramic. This condition limits the integrity of the joints to 600°C in practical use . Common matching criteria relate only to the static case, and do not apply during processing and in actual service. Computed dynamic thermal mismatch strains and stresses may be more than five times greater than the corresponding static mismatch values. Dynamic mismatch, as opposed to static mismatch, usually leads to failure of the metal-ceramic joint. To minimize these dynamic mismatch stresses and strains, we have developed a method in which microjoints in parallel are graded laterally, or from the center to the periphery of the bonding interface region, in thermal conductivity, stiffness, and/or thermal expansion coefficient. All present graded seals (microjoints in series) vary only the thermal expansion coefficient axially, or from the metal to the ceramic, to reduce static mismatch stresses and strains. Our laterally graded seals are useful for mounting fragile silicon, GaAs, or diamond chips and 1-2-3 ceramic superconductors onto metal substrates for improved bond strength, heat dissipation, and device performance.

Metal - Ceramic Joining
Edited by P. Kumar and V. A. Greenhut
The Minerals, Metals & Materials Society, 1991

Introduction

The joining of ceramics to metals has always been very difficult. Reliable ceramic joints for high temperature uses do not exist anywhere in the world.

Japan has a well-coordinated intensive effort on metal-ceramic joining. According to Prof. T. Suga of the University of Tokyo in his 1989 review paper on the "Future Outlook in Japan", they have already successfully used a ceramic turbocharger (NGK, Nissan), produced an all ceramic swirl chamber for diesel engines (Mazda, NGK), and prototyped a ceramic turbomolecular pump (Mitsubishi and Japan Atomic Energy Research Institute). But their best ceramic joints to special "matching" alloys have a limiting useful temperature of 600°C. Their bond strength decreases rapidly with temperature because the reaction products in their bonded regions become weak and brittle under thermal stresses (1). Japan considers the improvement of the thermal resistance of their joints to be an urgent task.

The European effort, mainly in France and Germany, has been even less successful. Germany did not reach their goal after the first ten-year (1974-1983) program and its follow-up in 1983-1986. Their present program (1985-1994) concentrates on the achievement of reproducible mechanical properties and component reliability (2). The U.S. Department of Energy supports much of the development of U.S. ceramic joining technology in the automotive industry. This work, however, has continued for a long time and at great expense.

Many problems exist in metal-ceramic joining. The science of ceramic joining is not well understood. Most joining processes have been developed through trial and error based on a novel idea, equipment, process, or material. While these processes achieve some improvements, there are no major breakthroughs. The long-term chemical incompatibility of component materials is not considered. The bonding interface is overemphasized, while the interfacial region is not properly addressed. Also, the common matching criterion deals exclusively with the static case by assuming complete thermal equilibrium between the ceramic and metal, while neglecting the much more critical dynamic matching conditions. As a result, the reproducibility and reliability of the joint remains inadequate.

Factors in Joint Design

The factors in a metal-ceramic joint design should include the following:

1. Metal composition
2. Ceramic composition
3. Joint Failure Mode
   a. Failure on heating or cooling
   b. Power density
   c. Type, rate, dimensionality of heating or cooling
4. Parts Configuration
5. Parts Sizes
6. Thermal Design
7. Special requirements or operating conditions

Each important metal-ceramic joint and bonding process must be specially designed and developed. The factors in joint and process design are many and often complex, conflicting, or difficult to achieve, as shown above. Engineering compromises must be made. Much judgment is required in any design. The operational conditions and requirements for the National Aerospace Plane (NASP) are, for example, different from those of the diamond heat sinks or fusion reactors. A metal-ceramic bond designed for maximum mechanical strength is usually not the best for thermal conductance, which is critical in heat sinks. What is best for one application may even be precisely the worst for another, even for the same parts shapes, sizes, and material combinations. A joining method for many conditions may not be the best for any application.

Because of the complexity of the metal-ceramic bonding process and the many worldwide attempts to develop better bonding methods with the best available resources, simple trial and error approaches are unlikely to provide breakthroughs in ceramic-metal joining. The chance of major success from one or two innovative ideas, equipment, methods, or materials is therefore not great.

Rather, a methodic scientific analysis must be made with a system's approach from the chemical, mechanical, thermal, electronic, thermochemical, and mechanochemical (stress induced phase transition) viewpoints. Other factors often are also important. In a diamond-metal heat-sink bonding design, for example, mechanoelectronic effects must also be considered. This is because mismatch stresses cause carrier mobilities changes, leading to unstable or unpredictable electronic circuit characteristics. This type of analysis we have done for selected joint designs.

The requirements for the national aerospace plane (NASP) differ greatly from those of the diamond heat sinks or fusion reactors. In the NASP, the potential metals include titanium, steel, niobium, and aluminum, while the ceramics may involve Si_3N_4 and SiC. The predominant failure modes may relate to the engines being rapidly heated up to 1400 or 2500°C, or relate to the aerodynamically heated (to 3000°C) wings being instantly cooled in an ice storm. The generally one-dimensional heating or cooling from medium to high power have half times of minutes and seconds. But the best titanium-Si_3N_4 joint for the NASP turbine subjected to rapid heating should not be used for the wings of the same plane subjected to possible ice quenching failures.

In the fusion reactor program, building an economical reactor is the biggest challenge. Many experts consider the material problem insurmountable. The material failure modes in fusion reactors are very poorly understood. For example, severe erosion occurs and mysterious, hypersonic carbon (3400°C melting point) particles often explode out from the depth (3). Apparently, localized extremely high power concentrations, temperatures, and millisecond half heating times exist. Hemispherical or cylindrical heating appears likely. These carbon particles must have been instantly superheated to extreme temperatures for achieving hypersonic velocities and huge accelerations in the small reactor chamber. Hence, the refractory carbon materials must have special temperature-resistant carbon-metal joints. The graphite-metal joints made in 1990 at Sandia National Laboratories did not have adequate temperature resistance. If similarly superheated to extreme temperatures, refractory tungsten or platinum metals could also generate superheated molten, hypersonic metal particles of large masses.

In the design of diamond-metal joints for electronic heat sinks, the emphasis is also on high reliability. But the temperature (250-800°C) and power input is comparatively low, and the needed strength is minimal to hold the little mass of small diamond and tiny semiconductor chip. The most critical requirements here are long life, stable low thermal resistance, high, reproducible heat spreading capability, and minimum strain to the semiconductor chip to avoid carrier mobility changes (4). A totally different joint must therefore be developed, even though graphite and carbon in fusion reactors and diamond in the heat sinks have the same chemical composition.

Static vs. Dynamic Mismatch

Ceramics resist metallic wetting, requiring special surface metallizing before bonding onto metals for high strength uses. Different physical, chemical, and electrical filming methods may be considered for coating the ceramic. In a review paper in Physica Status Solidi, various chemical vapor deposition (CVD), evaporation, sputtering, and melt-growth techniques have been closely examined as to procedures, product qualities, defects formation, and yield (5).

Each of these and many other methods, have their unique advantages. Some, for example, are atomically precise. Others thoroughly clean the substrate for better adhesion or crystalline epitaxy which is necessary for most semiconductor devices. Some others produce splat cooling and superfine grains, with resultant increased Young's modulus. Still others operate at low temperatures to avoid unwanted thermal effects such as materials annealing, dopant redistribution, or evaporative segregation in, e.g., GaAs or HgCdTe.

However, all of these methods, and many others, fail to deal effectively with the critical problems of temperature resistance and thermomechanical mismatch. They all do little to lessen the rupturing mismatch stresses on the brittle ceramic material. Bond failures therefore result.

Statically, or under thermal equilibrium conditions, the thermal mismatch strain is proportional to the product of the thermal cycling range and the difference in thermal expansion coefficients between the ceramic and the metal substrate. This product, when multiplied by a Young's modulus, gives the static thermal mismatch stress. The Young's modulus may be some average modulus of the ceramic, the substrate, or both.

Actually, when heated or cooled during processing or service, the bonding members are not at the same temperature, particularly in a metal-ceramic joint. During heating of a metal-ceramic joint, the ceramic temperature lags behind that of the metal, often markedly so; while under cooling the opposite is true. This gives rise to different temperature profiles in the metal and ceramic at a particular time instant on heating or cooling. Dynamic mismatches then arise.

For a two-inch diameter end-to-end steel joint to Macor (Corning's machinable glass ceramic) cooled from 950°C to 20°C in air, the maximum temperature differential may reach 773°C about 1,000 seconds after the cooling. Also, the dynamic or transient thermal expansions coefficients and the resultant dynamic mismatch strains and stresses are not constants. These parameters depend highly on the material shapes, sizes, densities, surface properties, and heating or cooling conditions of the bonded members. The dynamic mismatch in thermal expansion coefficient for the joint may reach 29.6 ppm/^{o}C and the ratio of dynamic to static expansion coefficient mismatch may exceed 5.3 at a computable critical cooling time. The computed dynamic mismatch strain and stress for the same joint on cooling may reach maxima of 2% and 52,800 psi, respectively (6).

Dynamic mismatch stresses, not static ones, usually fracture the brazed metal-ceramic joint. Unless understood and properly controlled, these mismatches easily fail the brittle ceramic.

Graded Seals

Graded seals are used in making difficult ceramic-metal joints. Currently, all these graded seals are designed to overcome the static mismatch stresses and strains. These seals are exclusively of the serial type and grade only one property, i.e., thermal expansion coefficient. That is, the thermal expansion coefficient of the material of the bonding region between the metal and ceramic is monotonically changed from the metal toward the ceramic. Hundreds of United States and foreign patents deal with these serial grading methods.

A common graded seal has at least two or three layers of materials of changing or graded expansion coefficients placed between the metal and ceramic. A monotonically increasing or decreasing coefficient is achieved from the metal through these different grading layers to the ceramic. A three-layer graded seal thus has a bonding region with four dissimilar material joints in series. Many patented processes employ five or even more grading layers. During the bonding process, mass diffusion may occur between the metal, ceramic, and these intervening layers to achieve a more gradual or less abrupt grading for better results. The diffusion effectively increases the number of graded layers in the bonding region between the metal and ceramic, achieving the more efficient graded metal-ceramic microjoints in series.

As expected, the maximum temperature differential between the metal and ceramic always occurs at the central axis of end-to-end rod joint where the heat diffusion length is maximal. Several actions may be taken to minimize the dynamic mismatch stresses:

1. Increase the thermal conductivity of the material of the bonding region, particularly at the center where maximum dynamic mismatches occur, to minimize the dynamic mismatch strain and stress through reducing the temperature differential between the metal and ceramic;

2. Decrease the stiffness or Young's modulus of the bonding region material, particularly at the center, to reduce the mismatch stress even with the same mismatch strain;

3. Decrease the thermal expansion coefficient differential between the metal and ceramic; and

4. Any combination of the above three actions.

Our new laterally graded seals differ from the current, serially graded seals in several respects:

1. The new graded seals are designed not only to overcome static mismatch stresses, but to minimize the more critical dynamic mismatch stresses.

2. There are seven ways to design a new seal by grading the following three physical properties, taking one, two, or three at a time: expansion coefficient, a, thermal conductivity, k, and stiffness (or Young's modulus), E. Hence, the new graded seal design is more flexible.

3. The grading direction is not from the metal toward the ceramic, but is normal or lateral to this axial direction. The resultant, multiple cylindrical ring-shaped metal-ceramic joints, in combination, thus parallelly share the external load. Hence, we have the name graded metal-ceramic joints in parallel.

4. To improve the grading efficiency, solid diffusion may also be employed to vastly increase the number the bonding layers in the bonding region. Thus, the new graded seals consist of numerous graded metal-ceramic microjoints in parallel, not in series.

The following table summarizes the difference between the conventional serially graded and the new parallelly graded seals.

Table 1: Comparison of Graded Joints

Type	Conventional	New
Joints in	Series	Parallel
Grading	Axial	Lateral
To minimize	Static mismatch	Dynamic mismatch
Graded Property	a	k, E, and/or a
Thickness Control	Poor	Good
Flexibility	Poor	Good

Fig. 1 shows a comparison between the common, serially graded metal-ceramic joint and the new parallelly graded joint. In the former, the different bonding layers simply change the thermal expansion coefficient monotonically from the metal side to the ceramic. In the later, the different bonding layers in ring-shaped cylindrical tubes decrease the thermal conductivity but increase the stiffness and thermal expansion coefficient from the center to the periphery of the joint. At the center of the joint, the maximum dynamic mismatch occurs, requiring high conductivity, low stiffness, and small expansion coefficient difference between the bonding material and ceramic. That is, low expansive bonding material is preferable at this center. At the periphery, dynamic mismatch is minimal, thereby allowing relatively lower conductivity, more stiffness, and higher expansion differential relative to the fragile ceramic.

If a typical binary alloy, such as the eutectic type lead-tin alloy, is used in the design of new joint, the center may be the relatively pure tin or lead, which has both higher conductivity and is softer than the more alloyed, outer material. Hence, the high conductivity and softness requirements in the central region are almost always met. There is also a 50:50 chance that the thermal expansion coefficient of the pure component has a desirable lower or undesirable higher expansion differential relative to the ceramic. But since either side of the pure component lead or tin may be chosen, this chance is doubled to 100%. Favorable grading

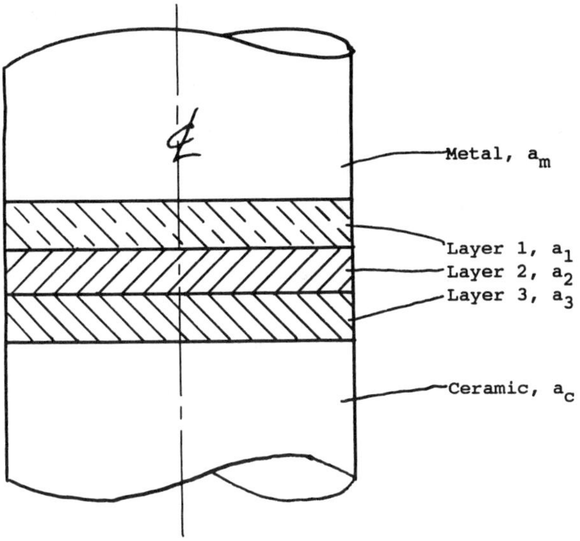

Fig. 1a: Serially or Axially Graded Joint

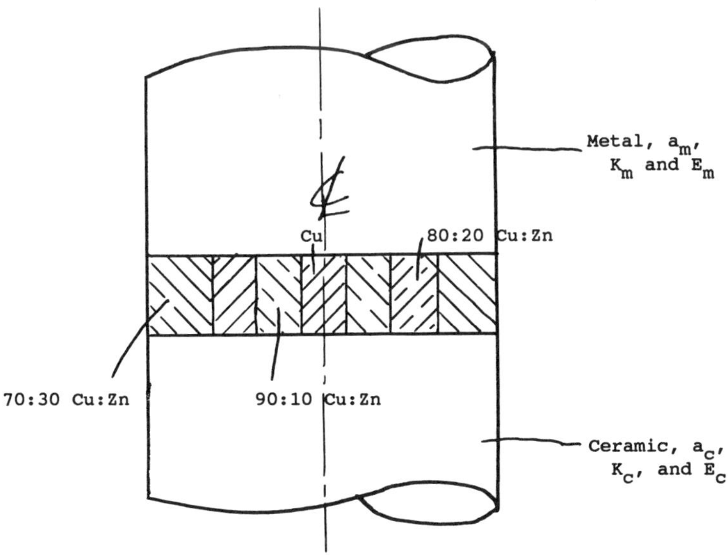

Fig. 1b: Parallelly or Laterally Graded Joint

Fig. 1: Graded Seals

in one or two of these physical properties may already be sufficient to provide useful practical results. But the probability of all the three physical properties (k, a, and E) being favorable is also good, even when other possible use requirements must also be considered.

The conventional, serially graded seals suffer in two respects:

1) The mechanically and chemically weakest layer, no matter how thin, controls the entire bond---Law of the Chain; and

2) During any processing or service, the possible existence of severe mismatch stresses between different layers or a single weak, brittle, or unstable phase in the entire relevant phase diagrams at any temperature, no matter how transient, can totally destroy the entire joint---Murphy's Law.

Note that the common orientation of all the microjoints in the new graded seal is parallel to the general direction of the principal applied loads and stresses. The critical area subjected to these stresses is not the entire cross-sectional area of the joint, as in the current, serially graded joints, but the different ring-shaped, circular material boundaries of infinitesimal thickness. Hence, the laws of Murphy and the chain do not apply. The first parallelly graded seal patent, U.S. No. 4,890,783, was issued to Lintel Technology in January 1990.

Uses of Metal-Ceramic Microjoints in Parallel

The new graded microjoints in parallel may be used in structural metal-ceramic joints, particularly for high-temperature applications. High temperature generates high dynamic mismatch stresses that may easily fracture the fragile ceramic, thus requiring special joint design.

In the electronic field, the new grading principle is useful in the manufacture of new preforms for bonding semiconducting silicon, diamond, gallium arsenide, indium phosphide, or silicon carbide chips onto metals. Here, the reduction of dynamic mismatch stresses and strains is highly desirable. In addition, the high conductivity in the central region allows better heat dissipation, which is important for many high-power, high-frequency, high-density circuits. The softness of the preform in the central region absorbs mismatch stresses on the semiconducting chip and stabilizes the circuit characteristics because of minimum carrier mobility changes due to these reduced stresses. The new ceramic superconducting films or devices may be similarly mounted onto metals with these preform also to enhance the bond strength, thermal management, and device performance.

Better understanding of the science of metal-ceramic bonding, such as these transient or dynamic matching phenomena, has enabled us to develop difficult-to-make metal-ceramic joints. For example, we have repeatedly made "unmatched" metal-ceramic structural joints that are exceptionally strong and thermomechanically shock-resistant. Our graphite-carbon steel (SAE 1010) joints withstood 950°C in slow cooling and 800°C in rapid quenching into ice water. In addition, these joints are almost indestructible mechanically as proven by repeated pounding with heavy metal objects such as (12-oz) hammers, confirming their high flexure test results (7). Our joints of square rods of alumina to SAE 1010 cold-rolled steel rods withstood a dozen 1,000°C rapid (within fraction of a second) quenchings into ice water. Additional mechanical shocks from 6-foot drop tests on carpeted, wood, or marble floors failed to break these joints.

Upon final forced fractures, the breaks occurred 1/4" away from the bonding interfaces, indicating defect-free joints. Our diamond-copper joints have been tested at $850^{O}C$ without failures [4].

In conclusion, a study of dynamic mismatch in metal-ceramic joints has revealed the maximum transient mismatch stresses and strains at critical moments during the heating or cooling of the joints. The magnitude of these calculated mismatches are much greater than those of the conventional static mismatches, which assume continuous uniform temperatures in the joints. The critical location of these potential failure sites has led to the development of a new lateral grading technique. This technique provides not only improved bond strength but better thermal management particularly important in electronic circuit bonding applications.

Prof. Franklin F. Wang of the State University of New York at Stony Brook and Prof. Bernard Gallois of Stevens Institute of Technology in New Jersey provided many valuable discussions on this and related topics in metal-ceramic bonding. This work was partly supported by U.S. Department of Energy Contract No. DE-AC02-86ER80382. Additional support was received from the New York State Science and Technology Foundation for the development of ceramic printed circuit boards.

Reference

1. T. Suga, "Current Research and Future Outlook in Japan," Designing Interfaces for Technological Applications: Ceramic-Ceramic and Ceramic-Metal Joining, ed. S. D. Peteves, (American Elseview, 1989), 235-245.

2. Cetyn, K., 1989, "Current Research on Ceramic Joining in Europe," Designing Interfaces for Technological Applications: Ceramic-Ceramic and Ceramic-Metal Joining, ed. S. D. Peteves, (American Elseview, 1989), 265-270.

3. Proc. of the 12th. Symposium on Fusion Engineering, V 1-2, IEEE Catalog No. 87CH2507-2.

4. C. H. Li, "Diamond Heat Sinks for Electronic Circuits," Final Report, ONR Contract No. N-00014-90-C-0143.

5. C. H. Li, "Epitaxial Growth of Silicon and Germanium," Physica Status Solidi, (1966), 3-45 and 419-442.

6. C. H. Li, "Improved Brazing Techniques for Machinable Glass Ceramics," Final Report to DOE Contract No. DE-AC02-86ER80382, 1986.

7. C. H. Li, "Improved Brazing Technique for Graphite," Final Report, DOE Contract No. DE-AC02-89ER80769 and NY State ERDA Contract No. 1407-ERER-ER-90, 1990.

REACTIVE METAL BRAZING OF ALUMINUM NITRIDE

R. K. Brow and R. E. Loehman

Sandia National Laboratories
Albuquerque, NM 87185

A. P. Tomsia

University of California at Berkeley
Berkeley, CA 94720

Abstract

The addition of titanium to eutectic braze compositions causes these alloys to readily wet and bond to AlN ceramics. Electron microscopic characterizations of the metal-ceramic interfaces reveal the presence of TiN, along with other Ti- and Al-containing phases. The formation of such interfacial reaction products is an additional thermodynamic driving force for the creation of useful metal-ceramic bonds.

The part of this work performed at Sandia National Labs was supported by the US Department of Energy under Contract No. DE-AC04-76DP00789.

Metal - Ceramic Joining
Edited by P. Kumar and V. A. Greenhut
The Minerals, Metals & Materials Society, 1991

Introduction

Aluminum nitride (AlN) is a candidate to replace aluminum oxide (Al_2O_3) in advanced microelectronic packages. AlN ceramics can have over ten times higher thermal conductivity than Al_2O_3-substrates currently in use and they possess a closer thermal expansion match to silicon, making them attractive materials for heat-sinks in high power devices [1].

Before this potential is realized, however, a number of technical issues must be resolved, including the development of a metal-cofiring technology, compatible thick- and thin-film metallization materials, and adequate joining techniques. The latter technology is particularly important for the production of hermetic packages that require metallized layers to attach package lids, to bond lead feedthroughs and pin grid arrays, and to attach chips to well metallizations. Techniques and materials currently used to metallize Al_2O_3 ceramics are not necessarily compatible with AlN because intergranular glassy phases, which bond to many thick and thin film materials but degrade thermal conductivity, are not present in the AlN materials. Furthermore, the interfacial chemistry responsible for wetting and bonding are distinctly different for nitrides and oxides.

The use of titanium active metal brazes is one promising technique for metallizing AlN ceramics. We have previously shown [2] that commercial Cu/Ag alloys containing titanium readily wet and react with AlN to form strongly adherent metallized layers. Transmission electron microscopic analyses of these interfaces reveal the presence of TiN and $(Ti,Cu,Al)_6N$ reaction products [3] that contribute to the formation of strong chemical bonds. In the present paper, we will review some of this previous work and some more recent brazing studies of Cu/Ti and Pd/Si/Ti alloys. We will show that titanium migrates to the AlN interface during brazing to wet the ceramic and to react and form strong chemical bonds. The interfacial reaction products are electronically and chemically compatible with both the ceramic and remaining metal, forming the type of intermediate phases necessary for successful metal-ceramic bonding.

Experimental Procedures

Most of the polished (<5μ-in) AlN substrates examined in this study were sintered with small amounts of Y_2O_3 as a densification aid; the manufacturer (Keramont Corp., Tucson, AZ) reports a room temperature thermal conductivity of 120-150 W/m-K The compositions of the reactive metal brazing alloys, both commercial (Wesgo Div., GTE Products Corp., Belmont, CA) as well as developmental, are listed in Table I. (The Cu/Ti and Pd/Si samples were made by Allied-Signal Corp, Parsippany, NJ.) Each alloy was in the form of thin (~0.1 mm) ribbons which were generally sandwiched between two clean AlN substrates (1 mm thick) to create the metallization reaction samples. In general, the commercial Cu/Ag brazes were bonded to AlN at 900-950°C, the 24Cu•76Ti alloy was bonded at 950°C, and the Pd/Si alloys were bonded at 950-1050°C. All reaction couples were heated for 30 minutes in argon. Graphite fixtures were used to hold and load (~200 psi) the sandwich configurations. Reacted samples were then cross-sectioned, polished, and the reaction interfaces were characterized by analytical scanning electron microscopy.

Table I: Compositions of the braze alloys (wt%)

Commercial Alloys*:		Research Alloys+:
Cusil:	28.0Cu•72.0Ag	76Cu•24Ti
Incusil:	27.0Cu•62.25Ag•9.5In•1.25Ti	94Pd•4Si•2Al
Cusil ABA:	33.5Cu•65.0Ag•1.5Ti	94Pd•4Si•2Ti
Ticusil:	26.7Cu•68.8Ag•4.5Ti	90Pd•4Si•6Ti

*from Wesgo Div., GTE Products Corp., Belmont, CA
+made by Metglas Prod. Dept., Allied-Signal Corp., Parsippany, NJ

In previous studies, we characterized the wetting behavior [2] of the Cu/Ag commercial alloys on AlN and used transmission electron microscopy [3] to examine their interfacial reaction products. The wetting samples were heated to 900°C at a rate of 100°C/min in a furnace with a p_{O_2} of ~10^{-16} atm and the contact angle of the molten drop on the AlN substrate was optically measured as a function of time. In addition, we studied the interfacial reaction chemistry of Ti-thin films and AlN using x-ray photoelectron spectroscopy; the details of these latter experiments can be also found in reference [2].

Results and Discussion

The results of the 900°C wetting angle studies of the Cu/Ag brazes on AlN [2] are reproduced in Figure 1. Alloys that do not contain Ti (i.e., Cusil, 28Cu•72Ag) do not wet AlN when heated above their liquidus temperatures, whereas alloys that do contain Ti readily wet AlN. Interestingly, alloys with greater Ti-contents have lower apparent wetting angles (defined as the angle reached after 60 minutes at temperature) and reach that minimum angle more rapidly, indicating that the initial Ti-activity affects both the thermochemistry of the interface and the wetting kinetics.

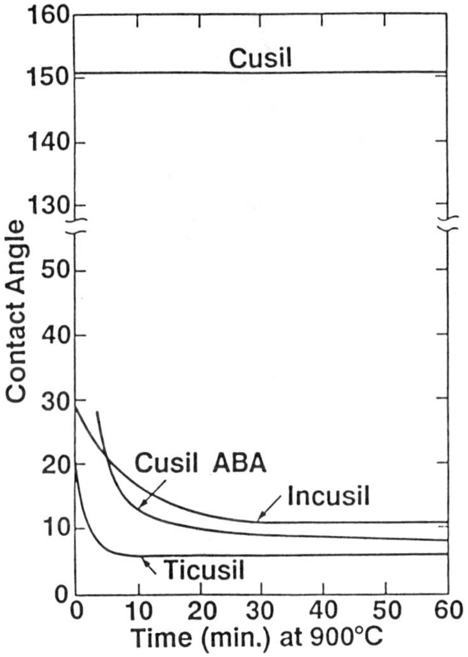

Figure 1: Wetting behavior of commercial brazing alloys on AlN substrates at 900°C [2]; alloy compositions are given in Table I.

Qualitatively similar results were obtained both for the 76Cu•24Ti alloy and for the Pd/Si alloys; those that contain Ti wet and spread on the AlN substrates, whereas those that do not contain Ti (e.g., 94Pd•4Si•2Al) ball-up and fall off from the AlN after cooling. We found that wetting by the Cu/Ti and Ti-doped Pd/Si alloys was sufficient to completely cover the surfaces of the AlN sandwiches and to fill 14 mil test vias that had been laser-drilled through the sample substrates.

Wetting behavior is usually described in terms of a balance in the interfacial energies according to the familiar Young-Dupre equation:

$$(\gamma_{sv} - \gamma_{sl}) = \gamma_{lv}\cos\theta \qquad (1)$$

where γ_{sv}, γ_{sl}, and γ_{lv} are, respectively, the solid-vapor, solid-liquid, and liquid-vapor interfacial energies, and θ is the equilibrium wetting angle. Wetting is defined as the reduction of γ_{sv} by the liquid and is driven by ($\gamma_{sv} - \gamma_{sl}$); when this term is positive (i.e., $\gamma_{sv} > \gamma_{sl}$), wetting will occur. Therefore, to promote wetting and spreading in nonreactive systems, γ_{sl} should be minimized.

The segregation of an alloy element to the solid-liquid interface is usually an indication of the reduction of γ_{sl}. For example, Figure 2 shows a backscattered electron image and a Ti x-ray map of the cross-section of a 76Cu•24Ti/AlN braze sample; titanium is clearly seen to segregate to the AlN-interfaces, leaving an essentially pure copper intermediate layer. The segregation of thinner Ti-layers can also be noted in the cross-sections of samples made with the Cu/Ag and Pd/Si alloys. (See reference [2] for micrographs of the former samples.)

While it is possible that segregation alone might sufficiently reduce γ_{sl} to produce adequate wetting, there is also clear evidence for the presence of interfacial reaction products that are not accounted for by the Young-Dupre equation. In fact, Yost and Romig [4] have shown that the free energy of formation of an interfacial reaction product (δG_r) will be an additional thermodynamic driving force for reducing θ; such an analysis yields a modified wetting equation [5]:

$$\gamma_{sv} - (\gamma_{sl} + \delta G_r/\delta A \cdot \delta t) \leq \gamma_{lv}\cos\theta \qquad (2)$$

where the interfacial reaction free energy is normalized to the change in the interfacial area (δA) and the reaction time (δt). The dynamic nature of this relationship makes the use of the inequality sign necessary [5]. According to equation (2), interfacial reactions with negative free energies will further reduce the influence of γ_{sv} and, if large enough, could promote wetting and spreading even in systems where $\gamma_{sv} < \gamma_{sl}$.

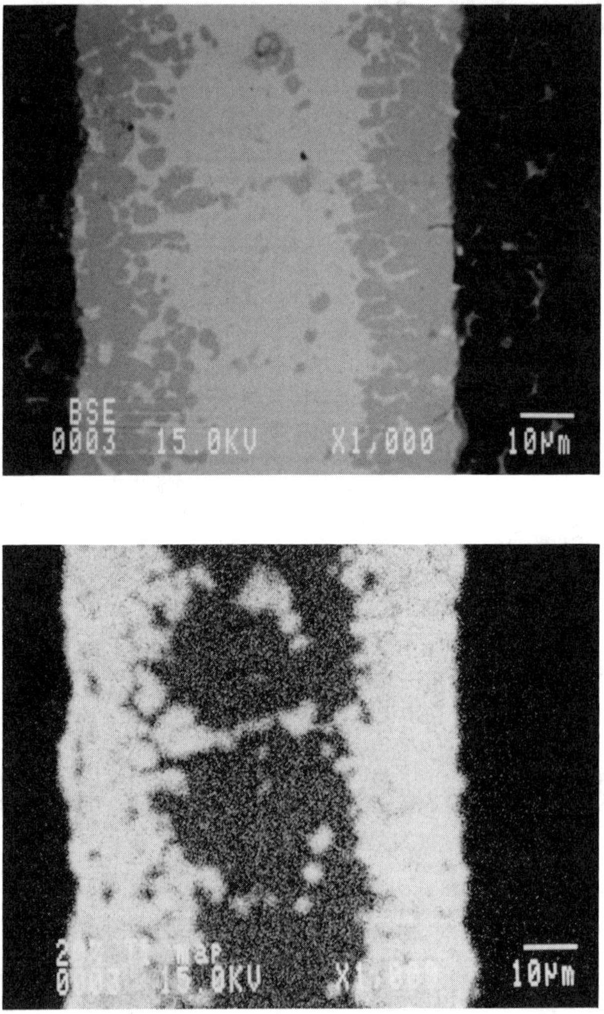

Figure 2: Backscattered electron micrograph (top) and Ti x-ray map (bottom) of an AlN/76Cu•24Ti reaction couple heated for 30 minutes at 950°C in argon; note that Ti has segregated to the AlN-interface.

We have studied the interfacial reactions that could further improve wetting, according to equation (2), by using x-ray photoelectron spectroscopy (XPS) to characterize the interfaces of heat-treated Ti-thin films deposited on AlN substrates. XPS is particularly useful for these studies because it is sensitive not only to different elements, but also to their different chemical environments. Figure 3 summarizes some of our thin film XPS results. The Ti*2p* (3a) and Al*2p* (3b) photoelectron spectra shown were obtained from a reaction couple (80 nm Ti on AlN) heated to 900°C for 30 min in argon and from several reference samples. The Ti/AlN couple was sputter-cleaned by an Ar-ion gun to remove an oxidized surface layer and to reveal the reaction products that include TiN (Ti*2p* binding energy ~455.0 eV [6,7]) and a metallic Al-phase (Al*2p* binding energy ~72.3 eV [7]).

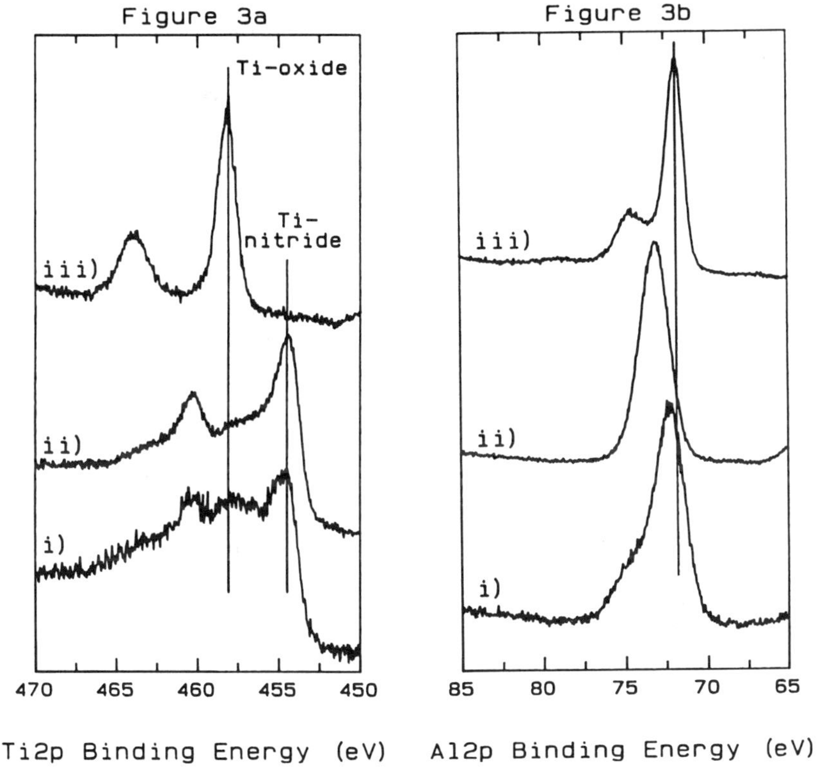

Figure 3: a) Ti*2p* photoelectron spectra from i) a sputter-cleaned Ti-thin film/AlN reaction couple and from ii) TiN and iii) TiO_2 reference samples; b) Al*2p* photoelectron spectra from i) the same Ti/AlN reaction couple, and from ii) AlN and iii) Al-metal reference samples.

The simplest reaction that is consistent with these XPS results is

$$Ti + AlN \rightarrow TiN + Al \tag{3}$$

Using tabulated free-energy data [8], this reaction has a δG_r of -8.8 kcal/mol at 1200 K and thus would provide the additional thermodynamic driving force for wetting given by equation (2).

In reality, the reaction products that form at Ti/AlN interfaces are more complex than those suggested by reaction (3). Beyers et al. [9] proposed a Ti-Al-N phase diagram predicting that $TiAl_3$ and TiN form when Ti reacts with AlN at 600°C. Norton et al. [10] reacted

TiH_2 films with AlN at 850°C and found Ti_2N and Ti_3Al at those interfaces. (It would be useful to do a careful study of the thermochemistry of the Ti-Al-N system, similar to what has been reported for Si-Ti-N [11].) Because the Al*2p* binding energies of these different Ti/Al intermetallics are virtually indistinguishable from that of Al and because the Ti*2p* binding energies of the different Ti/N phases are also quite similar, our XPS analyses could not adequately distinguish between those reaction products. However, the development of stable interfacial compounds, whatever their stoichiometries, still supports the suggestion that the addition of reactive components to a conventional braze improves its wetting as well as strengthens the interfacial bonding.

The presence of additional alloying elements, particularly Cu, further complicate the interfacial reactions. Previous TEM analyses of Cusil ABA/AlN interfaces [3] confirm that TiN is a reaction product that forms during conventional brazing operations. In addition, a second phase, identified as an η-nitride with a stoichiometry of $(Ti,Cu,Al)_6N$, forms between the TiN layer and the remaining Cu/Ag alloy (Figure 4). High resolution electron microscopy demonstrates that the AlN/TiN and TiN/η-nitride interfaces are abrupt, with no other crystalline or amorphous intervening phases [3]. These results are consistent with a gradual decrease in the Al- and N-activities, and a gradual increase in the Ti-activity as one moves from AlN across the reaction interface and into the braze alloy; such a gradual change in elemental activity produces the electronic and chemical gradient that is necessary for the formation of a strong metal-ceramic bond [5].

Figure 4: Transmission electron micrograph showing the reaction products that form at the AlN/Cusil ABA interface after 5 min. at 900°C; both TiN and the η-nitride phase can be detected.

While our studies have been primarily concerned with understanding the interfacial chemistry of reactive metal brazing of AlN, others have used this technology in packaging applications. For example, Miyashiro, et al [1] have attached Alloy 42 leadpins to AlN substrates using a Cu/Ag/Ti active metal braze and Asai et al. [12] developed a TiN-modified Mo paste that reacts with AlN for screening metallization patterns. Norton et al. have taken advantage of the reactivity of TiH_2 to braze AlN to copper and to another Fe-Ni alloy used to produce lead frames for ceramic dual-in-line packages [10]. The process conditions for these applications are compatible with those currently used to manufacture Al_2O_3-packages and so it seems that reactive metal brazing can be employed to produce AlN-based electronic components.

Summary

Our analyses suggest a sequence for the formation of strong metallization layers as a result of active metal brazing of AlN. When heated above their respective liquidus temperatures, the alloys melt and titanium segregates to the AlN interface to initially reduce the interfacial energy, γ_{sl}, enabling the alloy to wet the ceramic. Spreading and bonding are further enhanced as AlN grains dissolve into and react with the titanium. Reaction products, particularly TiN, nucleate and grow at the AlN surface to form phases that are chemically compatible with both the residual metal and the ceramic substrate, forming a strong metal-ceramic bond. On cooling, the ductility of the residual alloy is apparently sufficient to overcome expansion mismatches between the ceramic, the interfacial reaction products, and the metal, which gives excellent adherence of the metallized layer to AlN.

Acknowledgements

The authors wish to acknowledge the contributions of a number of colleagues, including R. H. Moore and R. G. Stone, who prepared many of the brazed samples, F. G. Yost, who supplied the Cu/Ti and Pd/Si alloys, and A. H. Carim, who did the TEM analyses.

References

1. F. Miyashiro, N. Iwase, A. Tsuge, F. Ueno, M. Nakahashi, and T. Takahashi, "High Thermal Conductivity Aluminum Nitride Ceramic Substrates and Packages," IEEE Trans. Comp. Hybr. Manuf. Technol., **13**[2] (1990), 313-319.

2. R. K. Brow, R. E. Loehman, A. P. Tomsia, and J. A. Pask, "Interface Interactions During Brazing of AlN," Advances in Ceramics, **26** (1989), 189-196.

3. A. H. Carim and R. E. Loehman, "Microstructure at the Interface between AlN and a Ag-Cu-Ti Braze Alloy," J. Mater. Res., **5**[7] (1990), 1520-1529.

4. F. G. Yost and A. D. Romig, Jr., "Thermodynamics of Wetting by Liquid Metals," MRS Proc., **108** (1988) 385-390.

5. J. A. Pask, "From Technology to the Science of Glass/Metal and Ceramic/Metal Sealing," Ceram. Bull., **66**[11] (1987), 1587-1592.

6. C. Ernsberger, J. Nickerson, A. E. Miller, and J. Moulder, "Angular Resolved X-ray Photoelectron Spectroscopy Study of Reactively Sputtered Titanium Nitride," J. Vac. Sci. Technol., **A3**[6] (1985), 2415-2418.

7. C. D. Wagner, et al., Handbook of X-Ray Photoelectron Spectroscopy, (Eden Prairie, MN: Perkin-Elmer Corp., 1979).

8. I. Barin and O. Knacke, Thermochemical Properties of Inorganic Substances, (Berlin: Springer-Verlag, 1977).

9. R. Beyers, R. Sinclair, and M. E. Thomas, "Phase Equilibria in Thin-Film Metallizations," J. Vac. Sci. Technol., **B2**[4] (1984), 781-784.

10. M. G. Norton, J. M. Kajda, and B. C. H. Steele, "Brazing of Aluminum Nitride Substrates," J. Mater. Res., **5**[10] (1990), 2172-2176.

11. S. Sambasivan and W. T. Petuskey, "Thermochemistry at SiC-Ti and Si_3N_4-Ti Interfaces," these *Proceedings*.

12. H. Asai, F. Ueno, N. Iwase, H. Sato, N. Mizunoya, T. Kimura, K. Endo, T. Takahashi, and Y. Sugiura, "Titanium Nitride-Molybdenum Metallizing Method for Aluminum Nitride," IEEE Trans. Comp. Hybr. Manuf. Technol., **13**[2] (1990), 457-461.

INTERFACIAL DEBONDING IN STAINLESS STEEL / GLASS-CERAMI

G.A. Knorovsky, R.K. Brow, R.D. Watkins and R.E. Loehman

Sandia National Laboratories
Albuquerque, NM 87185-5800

Abstract

Advanced pyrotechnic and electronic components can be fabricated from Ni-ba with hermetic seals to matched thermal expansion coefficient lithium-silicate (LSGC). Prior studies have characterized the interfacial reactions in these systen good chemical bonding. Similar interfacial reactions occur when LSGCs are series stainless steel except that these seals debond on cooling to room tempera found that Cr-depletion (from ~18 wt% to ~5wt%) from the steel causes an fc transition that expands the interfacial grains and decreases their thermal expans putting the LSGC into tension, causing the seal to fail. Linear elastic finite elem performed on a microcomputer suggest that the volume expansion associated transformation has a predominant effect compared with the difference in expansio

*This work was performed at Sandia National Laboratories, Albuquerque supported by the U.S. Department of Energy under contract number DE-AC04-7

Metal - Ceramic Joining
Edited by P. Kumar and V. A. Greenhut
The Minerals, Metals & Materials Society, 1991

Introduction and Background

lectrically-triggered pyrotechnic components are used to initiate events in many industrial and efense applications. Some familiar examples include air bag deployment in automobiles, arachute deployment for recovery or retardation of missile boosters and bombs, initiation of olid rocket boosters, and the explosive bolts used to rapidly open escape hatches in air and pacecraft. The key element in all these applications involves the rapid, correctly timed initiation f an event. Responding to a need for a highly energetic actuator in a confined space, searchers at Sandia National Laboratories developed a glass-ceramic/metal system [Ref 1] apable of providing high strength electrical feedthroughs in miniature closure welded explosive ackages. This technology offers an alternative to the complex and sometimes unreliable etallized and brazed alumina/Fe-29Ni-17Co system normally used in such applications. While omparable to ordinary glass/metal processing in the number of processing steps required, the ew technique provides much greater mechanical strength than ordinary matched or compression pe glass/metal seals. Also, the lithium silicate glass-ceramic (LSGC) provides improved hemical durability, and reduced susceptibility to cracking induced by welding. The original SGC/metal system used Alloy C-276 (C-276) feedthroughs, a lithium aluminosilicate glass, nd Alloy 718 (I-718) headers (see Table I for compositions). The metals (both Ni-base) were hosen for their corrosion resistance, weldability, and high strength. The LSGC was developed provide interfacial chemical bonding and a matched coefficient of thermal expansion (CTE). his latter is achieved by a multi step thermal treatment that nucleates a Li_3PO_4 phase (@ 00°C) and subsequently promotes the preferential growth (@ 820°C) of high CTE phases such s cristobalite and tridymite by epitaxy on the Li_3PO_4 particles [Ref 2]. Unfortunately, from the ewpoint of cost-effective manufacture, I-718 is difficult to machine, expensive, and difficult weld by the pulsed Nd:YAG laser weld process commonly used in manufacturing miniature omponents. A more satisfactory alternative, particularly where the high strength of the I-718 is ot necessary (as is usually the case) is the 304L grade of austenitic stainless steel. In what llows we will contrast the differing interfacial behaviors of the LSGC/I-718 and LSGC/304L stems, explain their differences using a phase diagram approach, and note how this affects the timate mechanical integrity of the metal/LSGC bond.

Results

SGC / I-718 Interfacial Behavior

s previously reported[Ref 3,4,5], interfacial reactions between the LSGC and I-718 destroy the

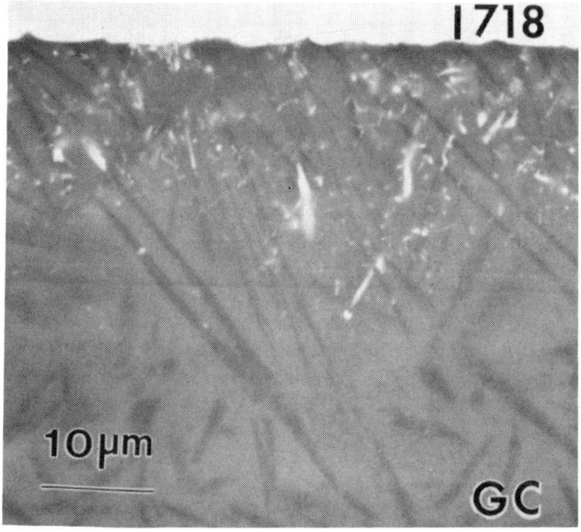

gure 1. Backscattered SEM image of LSGC/I-718 Interfacial Reaction Zone

Li_3PO_4 nucleation sites to a distance of several tens of microns into the LSGC. Cr rapidly diffuses from the I-718 into the LSGC, reacting with the Li_3PO_4 to form $Cr_{12}P_7$-type phosphides (the bright particles shown in Figure 1). This causes the formation of an interfacial reaction zone in the LSGC which, lacking the high CTE crystalline phases, has a CTE approximately 25% less than the desired value [Ref 6].

On cooling from processing temperatures, this reaction layer promotes residual stress development at the interface. Nevertheless, the strength of the LSGC/metal bond is such that when failure does occur (rarely), it occurs at the bulk LSGC/altered LSGC interface, and the system is found to be sucessful in practice.

LSGC / 304L Interfacial Behavior

Since 304L has a somewhat larger CTE than I-718 (~18 vs 14 x 10^{-6} °C^{-1}) a slight modification of the LSGC composition was required to obtain a greater tridymite phase content. This was done by substituting K_2O for Al_2O_3 in the base glass composition (see Table I). Similar to the LSGC/I-718 case, Cr is seen to have diffused into the LSGC (Figure 2). However, despite the increased Cr content, the crystallization is not significantly altered (Figure 3) and no interfacial reaction zone products are evident in the LSGC. Apparently, the Cr diffusion is not sufficient to form the phosphides encountered in the Ni-base I-718. However, despite the well-behaved crystallization behavior of the LSGC, on cooling after the crystallization step at 820°C the samples are found to debond at the LSGC/304L interface.

Figure 2. Cr x-ray map of an LSGC interfacial zone after debonding from 304L.

Discussion

Stable vs Unstable Metallic Behavior

In searching for a reason to explain this somewhat contradictory behavior, we eventually focussed on the metal side of the interface. We found that debonded 304L samples which had been left in ambient conditions showed rust on the surface previously covered by the LSGC, but not otherwise. Since "stainless" steel requires a minimum of ~12 wt/% Cr to exhibit its stainless behavior, it was clear that a very substantial decrease in interfacial Cr content had occurred. Further examination by x-ray diffraction (see Figure 4) proved that a body centered cubic phase was present in addition to the normal face centered cubic structure only where the LSGC had reacted with the metal substrate. The lattice constants of the two phases calculated from the diffraction peaks indicated that a 2% volume difference existed between the two phases at ambient temperature. Finally, Analytical Electron Microscopy (AEM) of metal foils taken parallel to and as close to the interface as possible had a Cr content of ~5 wt/%. The AEM also indicated that the morphology of the transformed phase most likely occurred via a diffusional reaction rather than a martensitic one, although calculations[Ref 7] based on the residual alloy content indicated that a martensitic reaction above ambient temperatures was possible. Clearly, a martensitic reaction or rapid diffusional transformation with a large volume change would be most detrimental to the LSGC/metal integrity if it happened at a temperature where the glass ceramic could not viscously flow, but even a slow diffusional transformation which the LSGC could accomodate would result in the formation of a bcc phase whose smaller CTE would be

Figure 3. Backscattered SEM image of the sample shown in Figure 2.

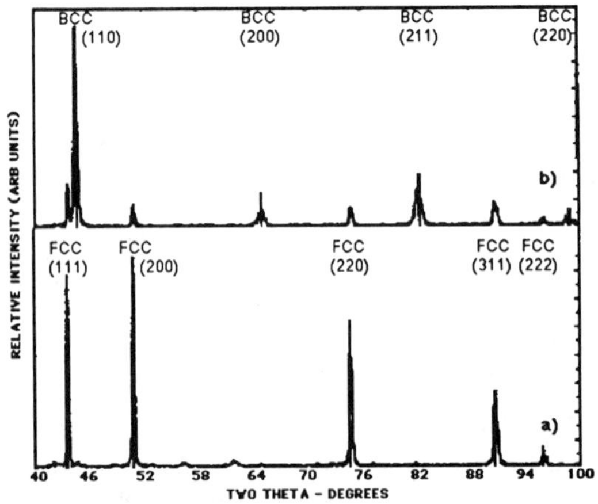

Figure 4. X-ray diffraction patterns from 304L a) before and b) after sealing to an LSGC

the cause of substantial residual stresses at the interface on cooling.

In both sealing systems, then, an intermediate layer forms that has a lower than desired CTE (in addition, the LSGC/304L system has a volume expansion). On cooling, since this layer contracts less than those on either side, it is put into compression, and consequently adjacent layers go into tension. In the LSGC/I-718 system, this means that the adjacent metal and bulk

LSGC must sustain tensile stresses. The metal is more tolerant of these stresses, hence when fracture is forced to occur, it will do so at the LSGC bulk/reaction layer interface. In the LSGC/304L system, the tensile stresses occur in the LSGC next to the bcc layer and in the underlying fcc matrix; debonding thus would occur at the LSGC/metal interface.

Finite Element Calculations

To better understand the relative roles of the difference in CTE and the volume expansion, a series of linear elastic finite element (FE) calculations were carried out on a microcomputer using commercial software (Cosmos/M by Structural Research Analysis Corp, Santa Monica, CA). These calculations were performed axisymmetrically in a simple cylinder-on-cylinder geometry with a "glue" layer as shown in Figure 5. Actually, the "glue" layer was broken into four parallel layers, as noted in Figure 5b. The same mesh geometry was used for all calculations, but the physical constants and thicknesses for the various layers (which are listed in Table II, and are not temperature dependent) were chosen to be consistent with experimental observations, with one exception. The thermal expansion coefficients used for all calculations were based on the LSGC/304L system. The actual CTE values for the LSGC/I-718 system are somewhat less (~30%), but we wished to determine the difference in stress resulting from the change in geometry of the interfacial layers without the complicating factor of the variation in CTE.

The specimens were cooled uniformly by 450°C to simulate cooling from the softening point of the glassy constituent of the LSGC. As may be seen in Figure 6, both systems show that the maximum principal stress (the most tensile stress) locations are to be found on either side of the reaction layer, with stress concentration at the surface. The size of the highly stressed region tends to be larger on the metal side, because it has stiffer elastic constants (by a factor of ~2.5 in Young's modulus). Without factoring in the volume expansion, the stress levels are somewhat lower for the 304L case, probably because the altered layer (layer 1) is thinner: 12.7 μm vs 63.5 μm. For the I-718 case the CTE difference was spread out over layers 2 & 3. Layer 2 is 25.4 μm thick and layer 3 is 37.5 μm; the latter layer's CTE was chosen halfway between the altered and normal values. Even if we had used the correct (~30% lower) CTE values for the I-718 case, the stresses generated for both cases would still be comparable, suggesting that they should behave similarly. When a ~1% volume expansion is added, Figure 7, the stress levels for the 304L case increase dramatically (a factor of ~3, note that the stress contours of Figure 7 are not identical to those of Figure 6). Some further calculations not reported here which allow plastic flow in the metal serve to reduce the stresses in both cases, but the 304L case still has substantially higher stresses. Based on these calculations it appears that the volume expansion effect is the most damaging factor causing debonding. Note that for reasons of limited space we are only showing the maximum principal stress; this has significance only for the brittle glass-

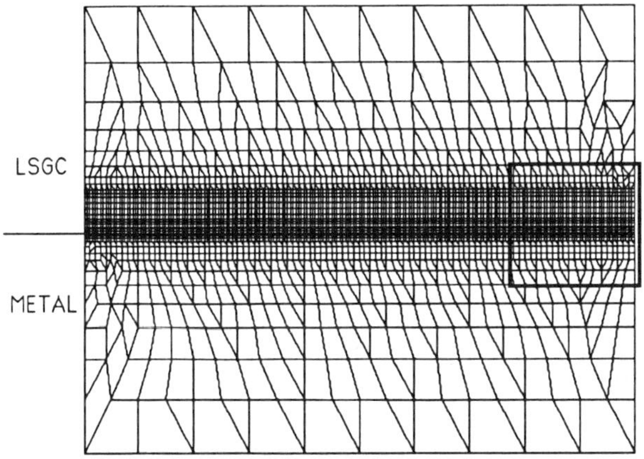

Figure 5a Overall mesh used; the left edge corresponds to the axis of symmetry, the cylinder has a radius of 2.5mm and overall thickness of 2 mm.

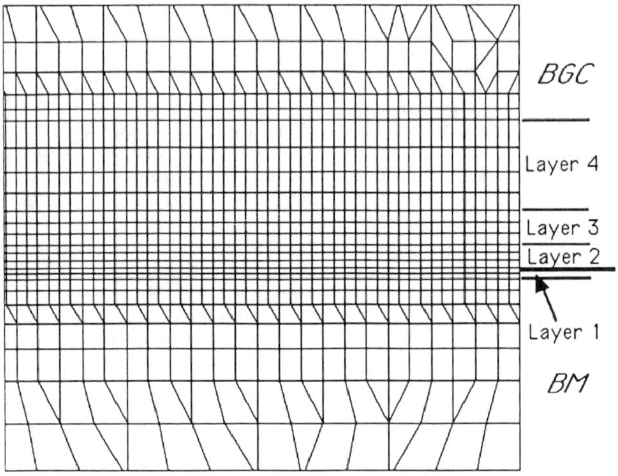

Figure 5b Detail of boxed area in Figure 5a. BM=bulk metal, Layer 1=transformed metal interfacial layer, Layer 2=most altered LSGC interfacial layer, Layer 3=less altered LSGC layer, Layer 4=unaltered LSGC layer (present to aid meshing only), BGC=bulk glass-ceramic.

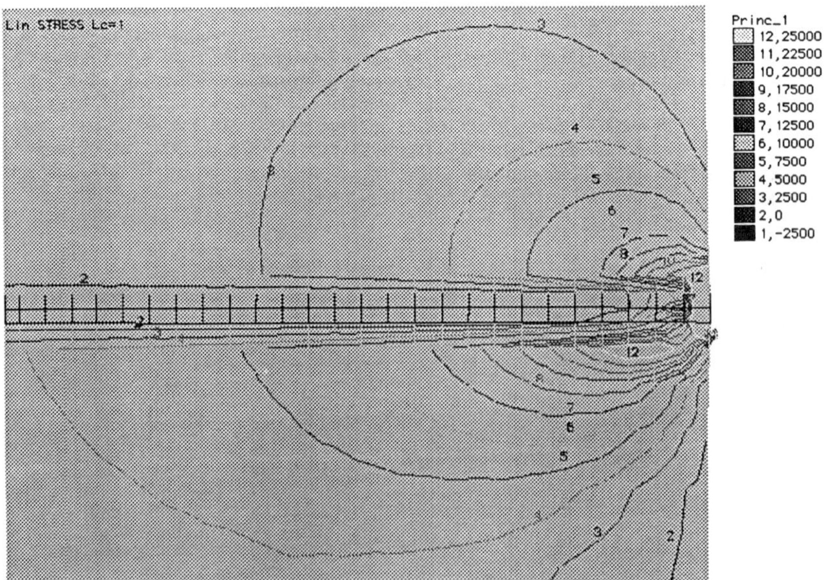

Figure 6a Maximum principal stress (in psi) in 304L/LSGC system on cooling; no volume expansion. Mesh elements shown are the transformed metal interfacial layer and are 12.7μm wide by 6.35μm high. This figure corresponds to a location at the outermost radius of the interface.

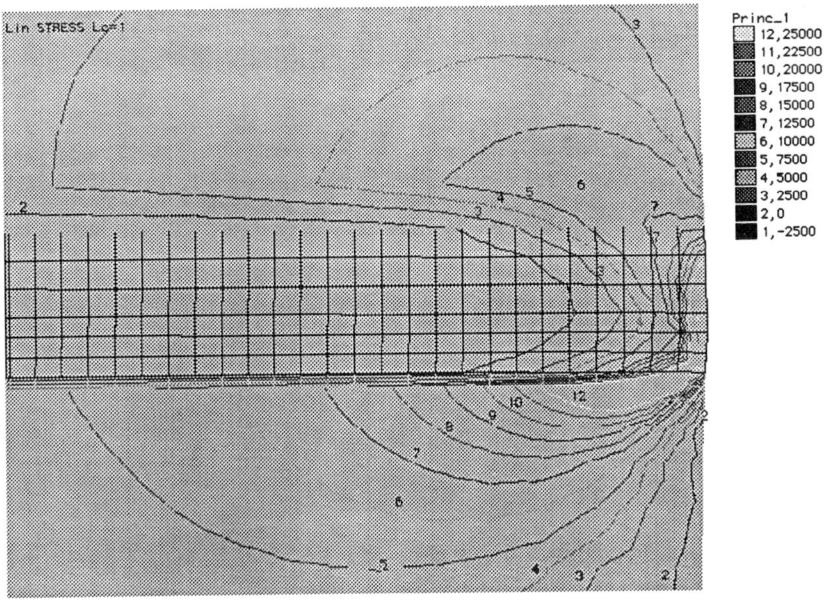

Figure 6b Maximum principal stress (in psi) in the I-718/LSGC system. Mesh elements shown are the most altered glass-ceramic layer (the three bottom element layers, 12.7 μm wide by 8.47 μm high) and less altered glass-ceramic layer (the three top element layers, 12.7 μm wide by 12.7 μm high). This figure corresponds to a location at the outermost radius of the interface.

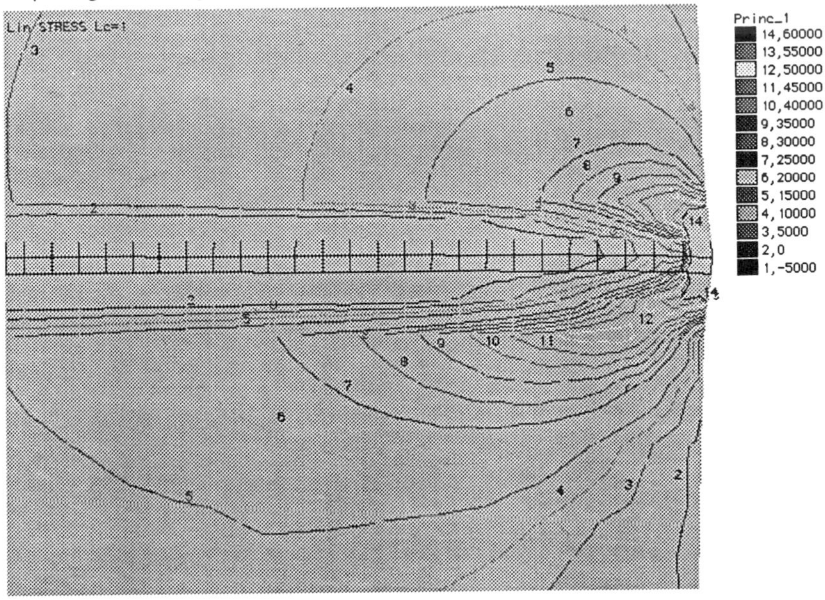

Figure 7 Maximum principal stress (in psi) in the 304L/LSGC system on cooling with 1% volume expansion of the bcc layer added. Mesh elements shown are the transformed metal interfacial layer and are 12.7μm wide by 6.35μm high. This figure corresponds to a location at the outermost radius of the interface.

ceramic. It would be more appropriate to plot the effective or Von Mises stress to show where plastic deformation in the metal would be expected.

Ternary to Binary Depletion Behavior

Understanding the differing behavior of the 304L compared with the I-718 is best achieved by looking at the binary Fe-Ni phase diagram Figure 8). The parent matrix phases (both fcc) of both 304 and I-718 can be thought of as predominantly ternary alloys of Fe-20 Cr-10 Ni and Ni-20 Cr-20 Fe, respectively (compositions given in wt/%, see Table 1 for more complete compositional information). If one removes all the Cr from both alloys, the residual matrix is

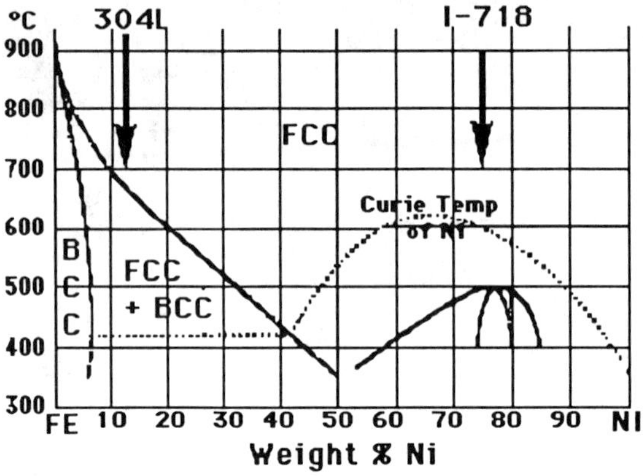

Figure 8. Fe-Ni binary phase diagram [Ref.9] showing the 304L and I-718 nominal compositions after Cr depletion.

essentially 87.5 Fe-12.5 Ni and 25 Fe-75 Ni, respectively. Looking at these compositions on Figure 8, (and ignoring the lower temperature phase changes which are not important because of their sluggishness) it is evident that upon cooling from the LSGC crystallization temperature, Cr depletion will take the fcc 304L into a region of the binary Fe-Ni diagram where the bcc phase appears. No such occurrence takes place at the Ni rich end of the diagram where Cr depletion takes the I-718.

Mitigation Strategies

Since an fcc-to-bcc transformation in the metallic substrate is apparently causing the debonding behavior, mitigative approaches can be suggested. However, whatever strategy is adopted, it should be compatible with the basic reason for trying to use 304L in the first place: i.e. cost-effectiveness. The approaches most evidently fall into two categories: 1) stabilize the fcc phase in the Cr depleted zone, or 2) prevent the Cr depletion. Implementing the first approach would require increasing the Ni or Mn surface content of the alloy by plating, or alternatively, by enriching the surface with an interstitial element such as nitrogen. Carbon and nitrogen are both very effective stabilizers of austenite; however, carbon promotes gas bubble formation at the interface, and also causes sensitization (a degradation of corrosion resistance). The second approach could take the form of Cr plating (increasing the Cr content in the bulk however would tend to promote bcc phase formation; 304L is actually situated in a relatively narrow window of fcc stability with respect to Cr content), or some other form of barrier to Cr diffusion. Preliminary results suggest that a preoxidation step which promotes the formation of an iron-rich oxide is helpful [Ref 8].

Conclusions

We have discussed bonding behavior in LSGC/I-718 and LSGC/304L systems, and have related their very different behavior to the underlying stability or instability of the metallic substrate to phase transformation caused by depletion of Cr. In particular, an fcc-to-bcc phase transformation in the LSGC/304L case results in debonding. In a more general conclusion, it may be stated that interfacial behavior affecting the success of LSGC/metal bonds can take place on both sides of or at the interface. Neither material can truly be thought "inert".

Acknowledgments

We are grateful to T. Headley and P. Hlava (both of SNL) for AEM and EPMA analyses, respectively, R. Tissot (also SNL) for x-ray analysis, and to A.P. Tomsia (U. C. Berkeley) for samples showing the results of preoxidation.

Table I Nominal Material Compositions (in wt %)

	SiO_2	Li_2O	Al_2O_3	K_2O	B_2O_3	P_2O_5
LSGC-A (I-718)	71.5	12.5	5	5	3	2.5
LSGC-B (304L)	75.5	13.5	0	5	3.5	2.5

	Fe	Cr	Ni	Mo	Nb	Si	Mn	P	others
I-718	18.5	18.5	52.5	3	5.1	0.2	0.2	0.01	0.9 Ti, 0.5 Al
304L	bal	19	10	-	-	1	2	0.02	0.03 C
C-276	5.5	15.5	bal	16	-	0.03	1	0.01	4 W, 0.35 V, 0.01 C

Table II Physical Constants Used in the FE Calculations

	CTE	Poisson's ratio	Young's Modulus
	Case 1: LSGC/304L, no volume expansion		
BM	18 x 10^{-6} °C^{-1}	0.3	2.07 x 10^5 MPa
Layer 1 (12.7 μm)	10.8	0.3	"
Layers 2-4 & BGC	18	0.25	1.725
	Case 2*: LSGC/I-718		
BM & Layer 1	18	0.3	2.07
Layer 2 (25.4 μm)	10.8	0.25	1.725
Layer 3 (38.1 μm)	14.4	0.25	1.725
Layer 4 & BGC	18	0.25	1.725

*Note that the actual CTE values should be ~30% lower; see text.

Case 3: LSGC/304L, with volume expansion

same as Case 1

BM=bulk metal, BGC=bulk glass-ceramic

References

1. H.L. McCollister, S.T. Reed, "Glass-Ceramic Seals to Inconel", U.S.Patent 4,414,282, Nov.8, 1983.

2. T.J. Headley, R.E. Loehman, "Crystallization of a Glass-Ceramic by Epitaxial Growth", J. Am. Ceram. Soc., 67, (1984), pp 620-625.

3. R.D. Watkins, R.E. Loehman, "Interfacial Reactions Between A Complex Lithium Silicate Glass-Ceramic and Inconel 718", Adv. Cer. Matls., 1, (1986), pp 77-80.

4. R.E. Loehman, "Processing and Interfacial Analysis of Glass-Ceramic to Metal Seals", in Technology of Glass, Ceramic, or Glass-Ceramic to Metal Joining, proc. Winter Ann. Mtg. of A.S.M.E., Boston, MA, Dec. 13-18, 1987, pp 39-45.

5. R.E. Loehman, S.C. Kunz, R.D. Watkins, "Reactions at Glass-Ceramic to Metal Interfaces", Ceram. Eng. and Sci. Proc., 7, (1986), pp 721-726.

6. S.C. Kunz, R.E. Loehman, "Thermal Expansion Mismatch Produced by Interfacial Reactions in Glass-Ceramic to Metal Seals", Adv. Cer. Matls., 2, (1987), pp 69-73.

7. G.H. Eichelman, F.C. Hull, Trans A.S.M., 45, (1953), pp.77-95.

8. A.P. Tomsia, private communication with R.K. Brow, 1989.

9. Metals Handbook Vol. 8, 8th Ed., ASM, Metals Park, OH, 1973, p 304.

EFFECT OF SUBSTRATE SURFACE ON BONDING OF Cu-AlN BY ACTIVE METAL THIN FILM AND GAS-METAL EUTECTIC METHODS

W.L. Chiang, L. A. B. Tessarotto*, V. A. Greenhut, and D. J. Shanefield

Department of Ceramics, College of Engineering,
Rutgers University
Piscataway, NJ 08855-0909

* on leave from Instituto de Estudos Avancados-IEAv/CTA, CP 6044, 12232-Sao Jose dos Campos, Sao Paulo, Brazil.

Abstract

Gas-metal eutectic (direct) bonded and active metal thin film solder bonded copper were investigated on AlN ceramics. It was found that copper-aluminum nitride bond strength depended on the particular source of AlN substrate and the pretreatment of the substrate surface. A satisfactory, reliable bond could be achieved by pre-oxidizing to form an alumina surface layer and/or sputtering a titanium and copper thin film layer onto the AlN substrate surface prior to bonding to copper. The strength of the bond in a peel test ranged from 15 N/cm to 78 N/cm depending upon substrate source and specific surface pretreatment. The strongest gas-metal eutectic bonds were similar in strength to the solder bond. The solder bonds, in turn, were limited by the strength of the soft solder itself.

Studies of the fracture mode for copper bonded to various pretreated AlN substrates showed that the failure interface changed with the surface oxide thickness. This may be related to a residual stress between the surface oxide and the AlN substrate resulting from thermal expansion mismatch when cooled from fabrication temperature.

Using the best preoxidation treatment, the bond strength with substrates from one supplier differed by a factor of three from the bond strength with substrates from another supplier in the case of the gas-metal eutectic method. A preliminary study showed that this might be attributed to different silicon content on the preoxidized AlN surface. This might alter the thermal expansion characteristics of the surface oxide layer and generate a smaller interfacial residual stress, which in turn results in a higher bond strength. A difference in calcium level may also enhance bond strength as suggested by a prior study of copper-alumina direct bonding.

A thermal shock test performed on soldered substrates indicated no appreciable deterioration in strength.

Metal - Ceramic Joining
Edited by P. Kumar and V. A. Greenhut
The Minerals, Metals & Materials Society, 1991

Introduction

Aluminum nitride (AlN) is an attractive candidate for high performance electronic packaging and mechanical applications. Its thermal conductivity is comparable to that of beryllium oxide, but AlN is not highly toxic. The thermal expansion coefficient of AlN is closely matched to that of silicon. Advanced mechanical applications are also developing.

One of the major problems that limit the use of AlN is bonding it to metal. In this paper, AlN was joined to copper using gas-metal eutectic and active metal solder methods. The effect of AlN substrate source and surface pretreatment (oxidation and/or Ti-Cu thin film) are reported.

Gas-metal eutectic bonding

The gas-metal eutectic is a direct bonding method. The basic idea of this technique is to utilize the eutectic reaction formed between copper and oxygen.[1,2] A schematic of this process and part of the Cu-O phase diagram are shown in Fig. 1. The copper is placed in contact with ceramic, and the whole assembly is heated to a temperature between the Cu-O eutectic temperature and the melting point of copper in the presence of a controlled partial pressure of oxygen. A eutectic skin will form between copper and ceramic. If this liquid "skin" wets the ceramic surface, it may bond the copper and the ceramic with an interfacial phase such as copper aluminate. A strong bond may result after the sample is cooled down. This method has been applied successfully in bonding copper to various oxide ceramics, especially alumina.[3] It provides low thermal resistance, which satisfies the increasing demand for high thermal dissipation capability for various applications. This method can not be used for joining copper to AlN unless the AlN surface has been pretreated. This is because AlN has lower surface energy than that of most oxide ceramics, so the eutectic melt does not wet the AlN surface. In the present research, several oxidizing pretreatments were studied for two of three commercially available AlN substrates. Peel strength and the mode of failure were determined.

Solder bonding to active metal thin films

Sputtering thin films of titanium and copper or other active metal has been used to modify a ceramic substrate, thereby permitting a conventional soldering process to join the ceramic to a metal member. It is well known that titanium provides excellent bonding on alumina. It has been explained that chemical reactions can occur between these materials which promote high adherence by substitution of Al^{3+} by Ti^{2+} in the alumina surface.[4,5] The schematic of this process is shown in Fig. 2.

For electronic applications a conductive copper coating is desirable onto the AlN substrate. However, copper sputtering adherence on AlN was found to be very weak. Titanium, as indicated, forms a very adherent bond, but it reacts strongly with oxygen. This means that an oxidized titanium layer forms immediately during conventional soldering, preventing solder wetting. A titanium sputtered coating of AlN, must be protected with a less reactive metal such as copper. Titanium and copper were sputtered onto cleaned AlN substrates using pure argon for the ionizing discharge.

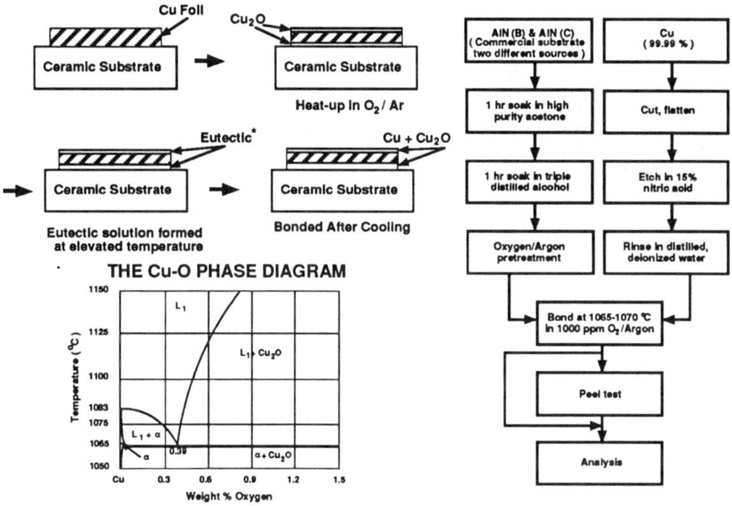

(a) (b)

Figure 1- (a) Schematic of the gas-metal eutectic method, and copper-oxygen phase diagram. (b) Procedure for copper-AlN direct bonding.

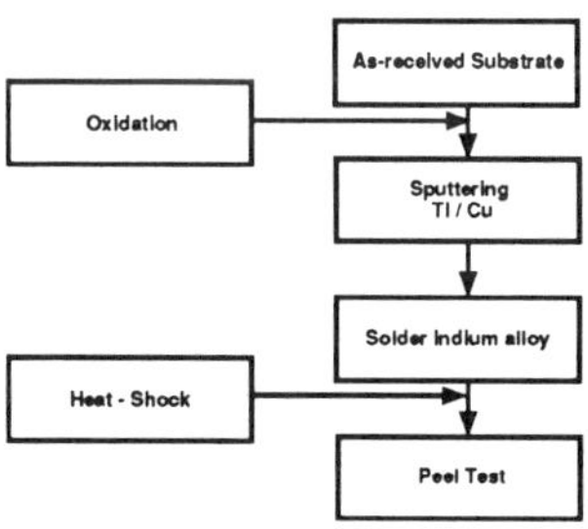

Figure 2- Procedure for active metal thin film method.

This was done without breaking the vacuum of the sputtering chamber so that the titanium would not be oxidized. The sputtered samples are then bonded to copper foil using a low temperature indium alloy solder. This process was applied to both oxidized and as-received AlN substrates from three different suppliers. Peel tests were used to evaluate bond strength of as-fabricated bonds as well as those subjected to five cycles of thermal shock.

Materials and Experimental Design

Copper foil (99.99%) was obtained from Electronic Space Products International Inc., Agoura Hill, CA. Commercial AlN substrates, designated "B," 1X1X0.025 inch (0.025x0.025x0.0006 m), "C," 1X1X0.040 inch (0.025x0.025x0.001 m) and "D," 1x1x0.025 inch (0.025x0.025x0.0006 m), were obtained from three different suppliers.

The surface roughness of all AlN substrates was about 0.5 μm. These AlN substrates were cleaned and subjected to surface pretreatment and bonding. The bonded samples were subjected to a 90° peel test and the bond strength was defined as average load per unit copper width.[3,6] A more complete description of the peel test may be found in the article "Peel Adhesion Bond Strength of Direct Bonded Copper-Alumina as Affected by Alumina Sintering Aids." in this volume.

A thermal shock test was performed by immediately plunging from a cold acetone liquid bath at -40°C (5 min hold) to a hot glycerine liquid bath at 125°C (>5 min hold) and vice versa.[7] Both baths were controlled within ±2°C. The peel test indicated above was used to measure possible deterioration in bond strength.

The SEM analysis was performed on AMRAY 1400 microscope. A SEIMENS D500 diffractometer was used for X-Ray diffraction analysis [XRD] and Auger and XPS were carried out on a Perkin-Elmer 660 Scanning Auger Microprobe and Perkin-Elmer 5000 LS respectively.

Gas-metal eutectic bonding

The as-received substrates were cleaned and subjected to one of the following pre-oxidation treatments:

AlN Pre-oxidation	Condition
P1	1250°C, 0.1 hr in O_2
P2	1300°C, 1 hr in O_2
P3	1300°C, 1 hr in O_2 followed by 1500°C, 1 hr in Ar

After pretreatment, the AlN was bonded with precleaned copper at 1065-1070°C in a gas mixture of Ar containing 1000 ppm O_2. The bonded samples were furnace cooled.

Solder bonding to active metal thin films

Sputtering was performed with a S-Gun magnetron machine which consisted of two guns operating in RF and DC modes. The cleaned substrates were set on a bias planetary (only available for RF mode). Both titanium and copper cathodes supplied by E-VAC Products, CA, were two piece targets (primary and secondary), with 4N purity. Titanium was first sputtered onto a biased substrate in RF mode (forward power=250 W). As this step finished copper was sputtered in DC mode (320W). A 500 Å titanium thickness coating layer was chosen arbitrarily. A copper coating thickness study revealed that a layer less than 1800 Å dissolved into the solder. A 9800 Å sputtered copper layer was used in this work. After sputtering all samples were stored at room temperature for, at least 48 hours.

For soldering, low temperature indium alloy, (70%Sn, 18%Pb, 12%In), in wire form, diameter=0.03 inch (~0.00076 m) from Indalloy

Co. was used together with Kester soldering paste. The soldering process was performed on a hot plate on which was placed an auxiliary massive iron plate, to keep its temperature at 30±2°C above the indium melting point (162°C).[8] The substrate and the peel test Cu foil were previously fluxed and set on the plate. A half inch (~0.012 m) long wire indium alloy solder was placed on the hot substrate, tinning its surface. Afterward the copper foil was gently positioned on the solder. This process required less than 20 seconds.

Results and Discussion

A gas-eutectic bond cannot be achieved on the as-received AlN substrate. The surface was not wetted by the eutectic melt. Changing the surface chemistry of AlN substrate using preoxidation enhanced the wetting and adhesion. The effect of preoxidation on peel strength is given below.

Gas-metal eutectic bonding

The bond strengths for copper bonded to various preoxidized AlN substrates are summarized in Table 1.

It can be seen that the more oxidized AlN substrates showed lower bond strength. This is consistent with results reported by K. W. Petra and J. Gobrecht.[9] An annealing step was applied to the oxidized AlN substrate, treatment P3. The bond strength of these substrates with copper was not improved. As seen in Table 1, AlN from different sources resulted in very different bond strengths.

The SEM analysis of AlN substrate surfaces corresponding to various pretreatments as well as the as-received samples is shown in Fig. 3. It can be seen that an alumina layer was generated on the AlN surface by the oxidation procedure. This was also identified by XRD as shown in Fig. 4. These alumina surfaces show a spongy structure.

Table 1. Peel strength results-suppliers "B" and "C"

AlN Pre-oxidation	Peel Strength (N/cm)	
	AlN(B)	AlN(C)
P1: 1250°C/0.1 h	19 ± 6	56 ± 10
P2: 1300°C/1.0 h	15 ± 3	35 ± 5
P3: 1300°C/1.0 h + anneal	19 ± 1	35 ± 1

As the AlN was oxidized to a greater extent(P2), the alumina layers became thicker, and the loose structure was more exaggerated. The annealing step after oxidation(P3) effectively sintered this surface alumina, but during the annealing step small cracks were also generated on the surface. This may be due to surface shrinkage during sample cool-down. The effect of the alumina layer on bonding

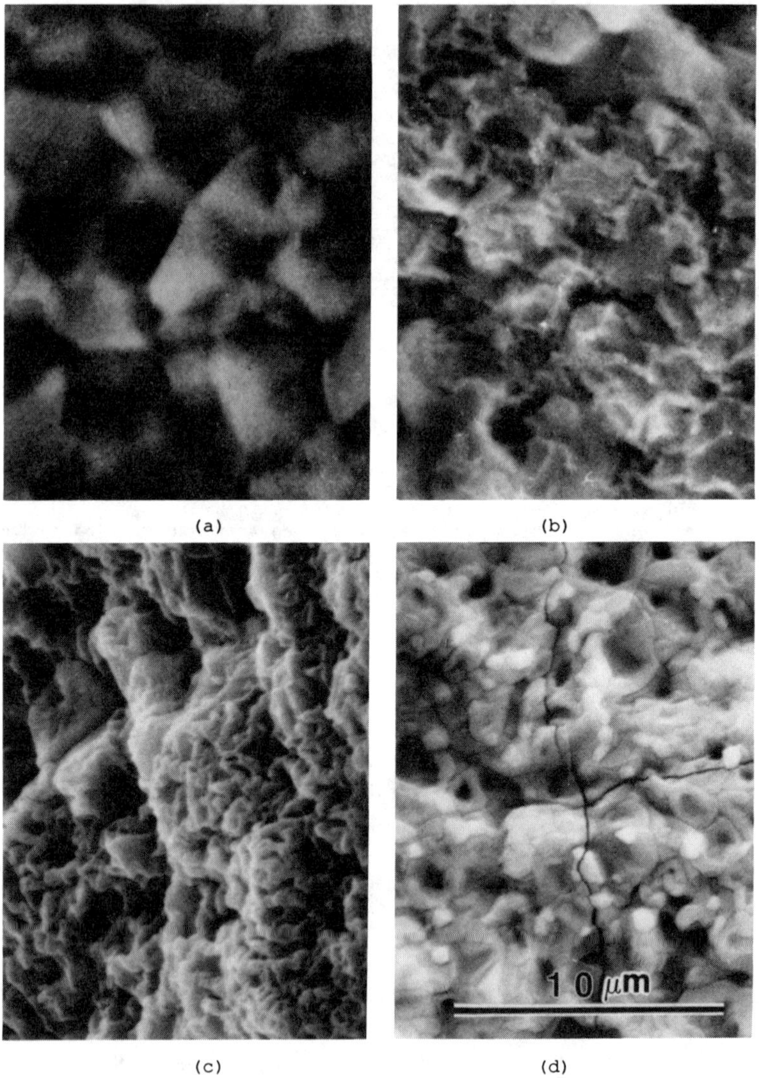

Figure 3- Scanning electron micrograph of substrate B surface after various pretreatments.
(a) As-received
(b) P1:1250°C 0.1 hr in O_2
(c) P2:1300°C 1 hr in O_2
(d) P3:1300°C 1 hr in O_2 followed by 1500°C 1 hr in Ar

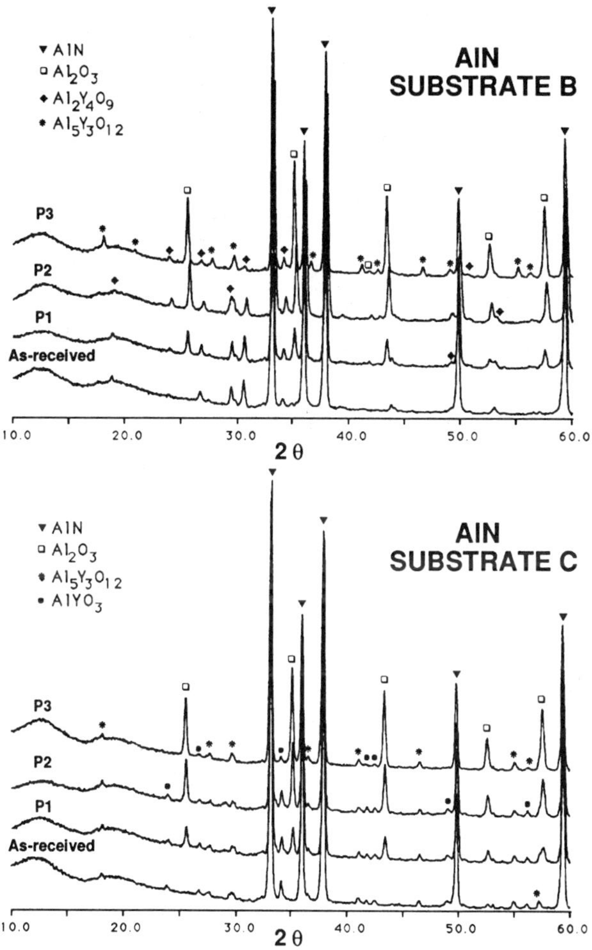

Figure 4- X-ray analysis of pretreated AlN substrates corresponding to the four conditions shown in Fig. 3.

may be explained by the observed porous structure, which may deteriorate the mechanical properties. An analysis of the failure mode was conducted in the cross section of the bond interface. Fig. 5 shows the SEM images of these cross sectioned samples. Before cutting the sample, the copper foil has been pulled a little bit to initiate a crack in order to identify the debonding interface. For copper bonded to AlN with a thin surface alumina layer(P1), the bond failure occurred in the $Cu\text{-}Cu_2O$ eutectic (Fig. 5a). However, when the alumina layer became thicker (P2 & P3) the failure mode changed.

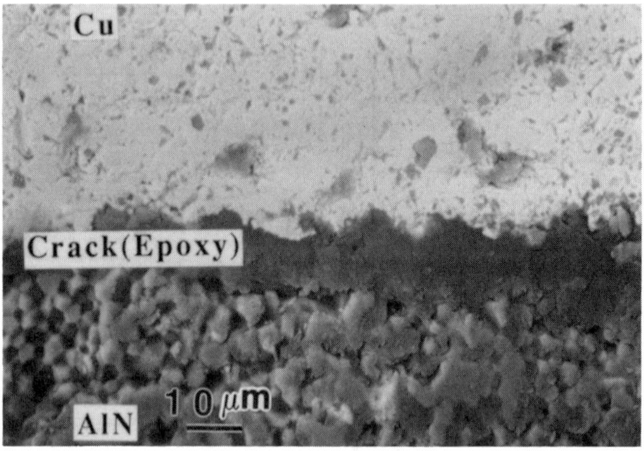

(a)

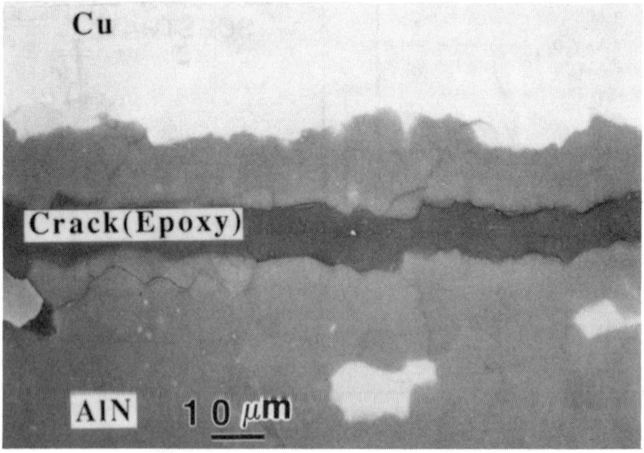

(b)

Figure 5- Scanning electron micrograph showing different fracture modes.
(a) Cu-AlN (1250°C 0.1 hr preoxidized in O_2)
(b) Cu-AlN (1300°C 1 hr preoxidized in O_2)

The crack propagated along the Al_2O_3/AlN interface which can be identified from the X-ray maps shown in Fig. 6. The observation of the change in failure mode may indicate that a residual stress exists between Al_2O_3 and AlN owing to thermal expansion mismatch. This is a factor which could deteriorate the bond strength. The thermal expansion coefficient of alumina ($9X10^{-6}$/°C) is almost double that of AlN ($5X10^{-6}$/C). The cracks produced on the alumina surface after high temperature annealing also suggest the existence of this stress causing cracks due to sintering shrinkage and differential contraction between AlN and Al_2O_3.

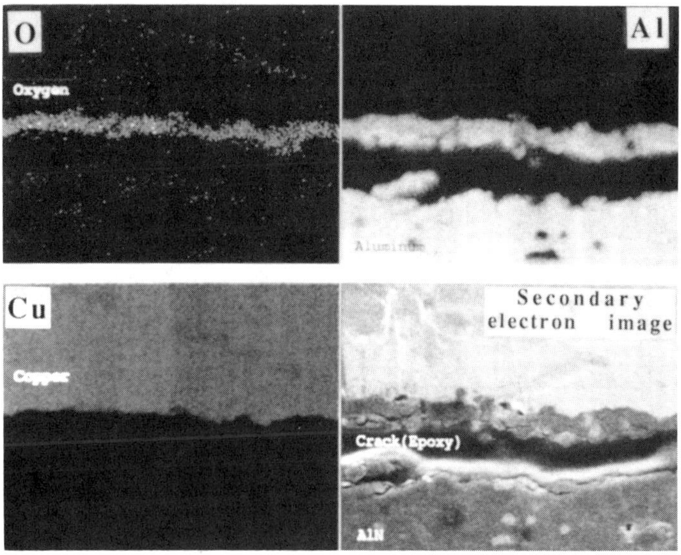

Figure 6- X-ray map of the sample shown in Fig. 4(b). The bond fracture occurred at the Al_2O_3/AlN interface.

The substantial difference in bond strengths to copper for AlN from different sources appears significant. A series of Auger analyses performed on substrates from different suppliers showed differences in surface composition. The result of an Auger analysis on the white rectangular area on the cross section of the bonded interface for copper bonded to substrate B and substrate C is shown in Fig. 7. Both AlN samples were preoxidized with O_2 at 1300°C for 1 hr (P2). On substrate C an obvious silicon peak was found, which diminished to a negligible level on substrate B. This is consistent with the results carried out on surfaces of as-received and cleaned substrates from the two suppliers by both Auger and XPS analysis. They are shown in Fig. 8 and Table 2, respectively. The higher silicon content in substrate C might explain the high bond strength when it is bonded to copper. The silicon present on the substrate surface could react with an alumina generated during the preoxidation process and form alumina/mullite solid solution. The thermal

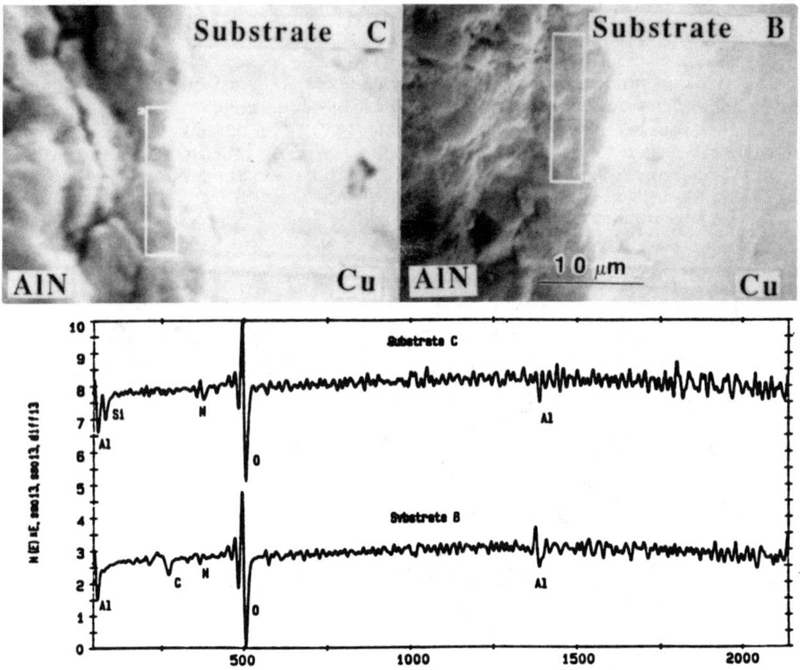

Figure 7- Auger analysis on the white rectangular area for substrate B and substrate C respectively. Both substrates were preoxidized at 1300°C for 1 hr in O_2 prior to bonding with copper.

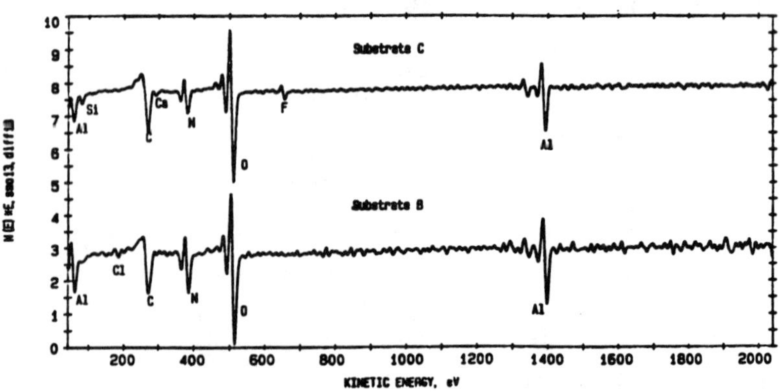

Figure 8- Auger analysis on the as-received and cleaned surface for both substrate B and substrate C.

expansion coefficient of mullite ($5X10^{-6}/°C$) is very close to that of AlN, so the formation of this phase might reduce the thermal expansion mismatch between surface oxide and bulk AlN, and result in a stronger bond. There is also a higher calcium content in substrate C and prior studies on copper-alumina bonding showed the Ca can enhance bonding.[3] A similar effect may occur here. This is the subject of continuing research.

Table 2. XPS analysis of substrate surfaces correspond to the samples in Fig.7 (excludes carbon)

Component	AlN (B)	AlN(C)
Al	38.4%	32.8%
O	39.4%	47.5%
N	19.8%	15.1%
Y	1.4%	1.0%
Ca	0.2%	0.7%
Si	0.8%	2.9%

Solder bonding to active metal thin films

The copper bonded peel strength data for as-received, preoxidized and thermal shock are summarized in Table 3. The oxidation condition used to pre-treat the substrate surface prior to sputtering, corresponds to the highest peel strength achieved in the gas-metal eutectic bonding method (P2).

Table 3. Peel strength results for substrates from suppliers B, C and D; all substrates are sputter coated with 500 Å of titanium and 9800 Å of copper.

	Peel strength (N/cm)		
Condition (Sputtered Ti-Cu)	AlN (B)	AlN(C)	AlN(D)
as-received AlN	78 ± 9	50 ± 25	78 ± 11
as-received AlN, thermal shock	52 ± 19	62 ± 14	76 ± 1
pre-oxidized (P1)	74 ± 5	70 ± 9	67 ± 6
pre-oxidized, thermal shock	80 ± 6	78 ± 8	76 ± 6

These results are related to the highest and constant load region of the displayed profile. Fig. 9 shows a typical view of the substrate soldered to Cu foil; Fig. 10 shows the AlN substrate coated with titanium and copper sputtered and the X-ray analysis of that region.

According to X-Ray analysis of tested samples, indium alloy solder was detected on the highest load peeled region, on both Cu

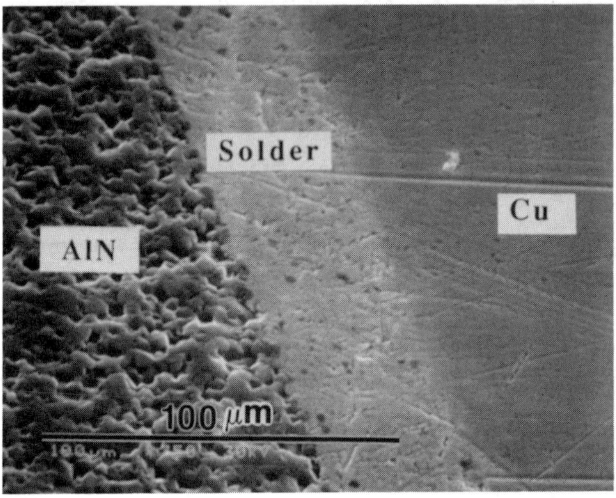

Figure 9- Titanium-copper sputtered onto AlN substrate, then soldered to copper foil, using indium alloy.

foil and substrate. This means that the maximum solder strength was achieved, and probably interface adherence itself might be greater. Observing the averaged strength results, in fact their values are very close to each other. Nevertheless a substantial reduction of random error is observed in the substrate C, pre-treated. It is clear that the standard deviation related to the pretreated substrates decreased to values below 9 N/cm. A remarkable comment can be made regarding the failure mode: observing the as-received substrate after test, large discontinuity areas could be seen in the metallized layer. The same is not observed in the preoxidized substrates where only a little material is pulled off. Additional facts must be emphasized: the peel strength profile behavior from those pretreated substrates became smoother in a large region than those for the as-received profile. Probably the substrate surface composition changes the adherence, but it has not been investigated yet.

The thin film metallization on AlN exhibits good resistance to thermal shock comparing the peel strength average of those three samples not submitted to the thermal shock with the average of those submitted to this test.

Conclusion

The bond strength of copper bonded to AlN depended on the particular AlN substrate used, the pretreatment of the substrate and the oxide thickness.

The surface oxide thickness of the preoxidized AlN substrate will affect the failure mode of a Cu-AlN bond. This may be explained by a residual stress at the oxide/AlN interface due to thermal expansion mismatch. The substantially different strength for copper

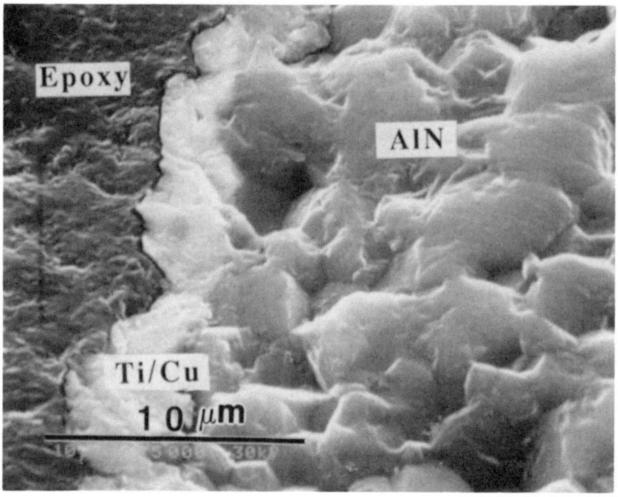

(a)

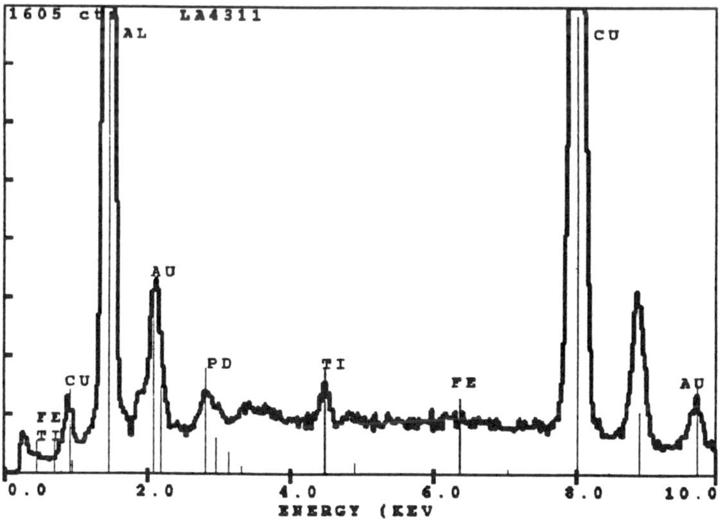

(b)

Figure 10- (a) AlN-titanium-copper interface region. The white layer corresponds to copper sputtered onto titanium.
(b) X-ray analysis performed on the above interface.

bonded to AlN substrates from different sources could be related to different silicon content of the AlN substrates. The presence of silicon may reduce the thermal expansion mismatch between the surface oxide layer and bulk AlN. The different residual stresses generated at the interface thus accounts for the difference in bond strength. This result suggests that the surface composition of the AlN substrate appears to be an important factor on Cu-AlN joining. More thorough studies are need for a complete understanding of these effects.

A soldering method achieved a reliable joint after oxidizing and metallizing the as-received AlN material, even though measurement was limited by using a soft solder. A 65 N/cm average in the peel strength of oxidized substrate, may be explained by high affinity of titanium to Al_2O_3 rather than AlN. No appreciable deterioration in strength was observed in the average peel strength between those substrates subjected to thermal shock compared to those not subjected.

Acknowledgements

The authors thank Jon Oliver for SEM work and Lois Johnson for XRD analysis, and L. Salvati and R. L. Moore from Perkin-Elmer Corp. for Auger and XPS analysis. Thanks also go to John Holowczak, Shih-Pong Young and David Lloyd for helpful discussions.

The authors acknowledge the support of the Center for Ceramic Research at Rutgers University in sponsoring this project. The support of the New Jersey Commission on Science and Technology and our corporate sponsors is gratefully appreciated.

The support of the CNPq of Brazil for L.A.B. Tessorotto's scientific stay at Rutgers University is acknowledged.

References

1. J. F. Burgess and C. A. Neugebaur, The Direct Bonding of Metals with a Gas-Metal Eutectic, U.S. Patent, No. 3, 744,120, (1973).

2. J. F. Dickson, "Direct Bond Copper Technology: Material, Methods, Applications," Int. J. Hybrid Microelec., 5(2), (1982), 103-109.

3. J. E. Holowczak, V. A. Greenhut and D. J. Shanefield, "The Effect of Alumina Composition on Interfacial Chemistry and Strength of Direct Bonded Copper-Alumina", Cer. Eng. Sci. Proc., Vol. 10 (9-10), (1989), 1283-1294.

4. W. H. Kohl, Materials and Techniques for Election Tubes, (New York, NY, Reinhold Publishing Co, 1960).

5. C. I. Helgesson, Ceramic-to-Metal Bonding, (Cambridge, MA, 1980).

6. H. Mizuhara and K. Mally, "Ceramic to Metal Joining with Active Metal Filler Metal," Welding J., (10), (1985), 27-32.

7. "Vitrified Ceramic Materials for Electrical Applications," Electrical Insulating Materials, (Philadelphia, PA: The American Society for Testing Materials, 1986) D116-86, 19-24.

8. "Determining Solderability of Thick Film Conductors," Electronics (Philadelphia, PA: The American Society for Testing Materials, 1978) F357-78, 415-416.

9. K. W. Petra and J. Gobrechti, "Directly Bonded Copper Metallization of AlN Substrates for Power Hybrids," Res. Soc. Symp. Proc., Vol. 40, (1985), 399-404.

Subject Index

Author Index